T0320363

Nonlinear Electronics 1

Series Editor
Robert Baptist

Nonlinear Electronics 1

Nonlinear Dipoles, Harmonic Oscillators and Switching Circuits

Brahim Haraoubia

ELSEVIER

First published 2018 in Great Britain and the United States by ISTE Press Ltd and Elsevier Ltd

ISTE Press Ltd
27-37 St George's Road
London SW19 4EU
UK

www.iste.co.uk

Elsevier Ltd
The Boulevard, Langford Lane
Kidlington, Oxford, OX5 1GB
UK

www.elsevier.com

Notices

Knowledge and best practice in this field are constantly changing. As new research and experience broaden our understanding, changes in research methods, professional practices, or medical treatment may become necessary.

Practitioners and researchers must always rely on their own experience and knowledge in evaluating and using any information, methods, compounds, or experiments described herein. In using such information or methods they should be mindful of their own safety and the safety of others, including parties for whom they have a professional responsibility.

To the fullest extent of the law, neither the Publisher nor the authors, contributors, or editors, assume any liability for any injury and/or damage to persons or property as a matter of products liability, negligence or otherwise, or from any use or operation of any methods, products, instructions, or ideas contained in the material herein.

For information on all our publications visit our website at http://store.elsevier.com/

British Library Cataloguing-in-Publication Data
A CIP record for this book is available from the British Library
Library of Congress Cataloging in Publication Data
A catalog record for this book is available from the Library of Congress
ISBN 978-1-78548-300-4

Printed and bound in the UK and US

Contents

Preface

This book is devoted to students enrolled in Bachelor's or Master's degree programs or in engineering schools. By combining step-by-step approaches to theoretical aspects and practical exercises accompanied by solutions, this book facilitates the reader's knowledge assessment and understanding of the phenomena presented.

It is worth noting that circuit design and realization require knowledge on the behavior and interconnection of devices. Providing such knowledge is the aim of this book, which deals with certain aspects of the nonlinear domain, a very broad domain with a wide range of applications.

This book deals with several subjects that definitely require prior knowledge on analog electronics. However, the didactic approach to these subjects is gradual.

Subjects that are hardly covered by other books are presented here, for example, a comprehensive presentation of oscillators: low frequency, high frequency, amplitude and frequency stability, the nonlinear approach and the determination of oscillation amplitude. Several astable circuits are presented in order to illustrate their broad range and the various possibilities offered by wave generators in terms of design and realization.

This book is organized into six chapters and contains more than 40 exercises and solutions covering a large part of nonlinear electronic circuits.

Chapter 1 deals with nonlinear two-terminal devices. Chapters 2, 3 and 4 focus on the generation of sine wave signals, from low-frequency oscillators to high-frequency oscillators and quartz oscillators.

Chapters 5 and 6 are complementary, as they deal with the commutation and response of RC circuits to pulse input and astable circuits.

Each chapter is followed by a series of exercises and solutions aimed at helping the reader enhance his/her comprehension and knowledge on the subjects presented.

Brahim HARAOUBIA
June 2018

Nonlinear Two-terminal Devices

1.1. Introduction

The study of nonlinear devices involves complex calculation that requires approximations.

When dealing with nonlinear two-terminal devices, graphical methods based on the study of their characteristics are used. The most commonly used example is the voltage–current or current–voltage characteristic, as shown in Figure 1.1.

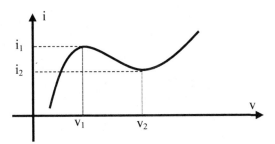

Figure 1.1. *Nonlinear "current–voltage" characteristic*

The "current–voltage" characteristic provides information on quantities such as static resistance in a well-defined point or dynamic resistance of nonlinear two-terminal devices.

It is possible to conduct a piecewise and approximation-based study of the characteristic in order to deduce the behavior of the nonlinear element.

1.2. Example of a nonlinear two-terminal device – the diode

It is worth recalling that a diode is a pn junction. It results from joining two semiconductors, one p-type and another n-type. The diode is a device with two poles, anode and cathode (Figure 1.2).

Figure 1.2. *Diode*

The diode conducts current in only one direction (forward direction), provided that the voltage applied between the anode and cathode exceeds the threshold voltage V_0.

This threshold voltage is imposed by the potential barrier that emerges when the p-doped and n-doped semiconductors are assembled.

When negative voltage is applied (($V_{anode} - V_{cathode}$) < 0), the diode is blocked and allows no current flow. In this case, the diode is said to be reverse biased.

When the voltage across a diode ranges from zero to the threshold voltage, the diode allows a very small current to flow through it. Given its importance for detection in the field of very high frequencies and also the specificity of the diode characteristic in this region, this aspect will be revisited.

The electric diagram of a diode is represented in Figure 1.3.

Figure 1.3. *Electric diagram of a diode*

$V_A - V_K > 0$: The diode is forward biased

$V_A - V_K < 0$: The diode is reverse biased

1.3. Characteristic of a diode

1.3.1. *Real diode*

The characteristic of a forward biased diode is described by the following equation:

$$I_D = I_{DS}(e^{\frac{nV_D}{V_T}} - 1)$$

where I_D is the current across the diode, n is a constant that depends on the diode-manufacturing process ($1 \leq n \leq 2$; generally, n =1), V_D is the voltage across the diode, I_{DS} is the reverse current and $V_T = 26\,\text{mV}$ at 300 K.

The representation of the diode characteristic $I_D = f(V_D)$ is shown by the plot in Figure 1.4.

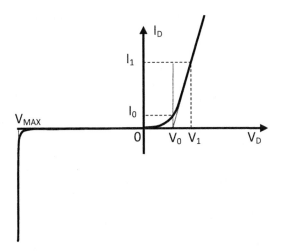

Figure 1.4. *Diode characteristic $I_D = f(V_D)$*

The above representation of the characteristic evidences at least four distinct regions as follows:

Region 1: $V_D > V_0$: the diode is forward biased. The characteristic is actually linear, and the diode is practically equivalent to its dynamic resistance "R_d". This part serves for detecting high-amplitude signals.

Region 2: $0 < V_D < V_0$: the diode characteristic can be assimilated to a curve resembling a parabola. In this region, the diode is used for the detection of weak signals. This is called quadratic law detection.

Region 3: $V_{MAX} < V_D < 0$: the diode is reverse biased. There is practically no current across it. In this case, the diode is equivalent to its reverse resistance "R_i". This resistance is very high, and in the ideal case, it is infinite.

Region 4: $|V_D| > |V_{Max}|$: the diode is destroyed under this condition.

Given the complexity of the real characteristic of the diode, far simpler models are used in order to facilitate the study, comprehension and also the possibility to rapidly design circuits, especially for the non-specialist. For this purpose, a certain number of approximations of the real characteristic of the diode can be adopted.

In this context, two approximations, which are in our opinion the most relevant, will be presented here.

1.3.2. Diode in first approximation

In the first approximation, a quite accurate rendering of the diode behavior (especially for high-amplitude signals) is represented by the characteristic schematically shown in Figure 1.5.

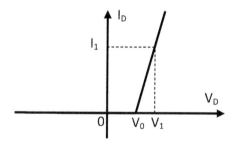

Figure 1.5. *1st approximation of a diode characteristic: $I_D = f(V_D)$*

It should be noted that in the given approximation the curvature evidenced on the characteristic when $0 < V_D < V_0$ is no longer present. Based on this characteristic, an equivalent diagram of the diode can be realized, which will allow for a very simple analysis of electronic circuits containing diodes.

The diagram in Figure 1.6 summarizes the behavior of the diode and its equivalent diagram in relation to the characteristic shown in Figure 1.5.

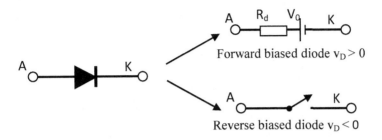

Figure 1.6. *Equivalent diagram of the diode in 1st approximation*

When the diode is forward biased ($V_D > 0$), it behaves as a voltage V_0 in series with a (dynamic) resistance R_d, which is the slope of the characteristic of the diode in first approximation:

$$R_d = \frac{\Delta V_D}{\Delta I_D}$$

When the diode is reverse polarized ($V_D < 0$), it is equivalent to its reverse resistance (R_i), which is considered infinite for the approximation in this case. The diode remains a nonlinear element despite the approximation made.

1.3.3. *Ideal diode – second approximation*

In this case, the diode is considered an ideal component. This is reflected by the following characteristics:

$$V_0 = 0; \qquad R_d = 0; \qquad R_i \to \infty$$

According to this approximation, the potential barrier V_0 created by the internal field is considered zero. Similarly, when the diode conducts, the resistance R_d that it opposes to current flow is considered zero and the resistance R_i that it presents when reverse biased is infinite.

This allows for deduction of the static characteristic related to an ideal diode (Figure 1.7).

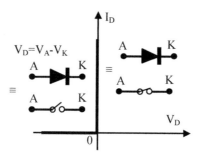

Figure 1.7. *Characteristic $I_D = f(V_D)$ of an ideal diode*

When the ideal diode is forward biased, it behaves as a short circuit or as a closed switch. When a weak voltage is applied ($V_D = V_A - V_K > 0$), the current across the diode is very strong. On the contrary, when the diode is reverse biased ($v_D = v_A - v_K < 0$), no current flows through it, regardless of the value of the reverse voltage applied. The diode can be assimilated to an open circuit.

1.4. Design of a thresholdless diode

An ideal diode has the following characteristics: threshold voltage $V_0 = 0$; dynamic resistance $R_d = 0$ and infinite reverse resistance R_i.

The practical approach to an ideal diode uses an operational amplifier circuit, as shown in Figure 1.8.

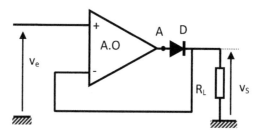

Figure 1.8. *Practical circuit for an ideal diode*

For the sake of clarity, the input voltage is assumed sinusoidal: $v_e = V_m.\sin(2\pi ft)$.

1.4.1. *Positive input voltage*

Initially, voltage v_S is zero. When voltage v_e is applied, the potential at the non-reverse input of the operational amplifier is higher than that at the reverse input. Therefore, the amplifier output is saturated (high state) and forces diode D to conduct, thus inducing a current across resistance R_L. The circuit equivalent to the one schematically represented in Figure 1.8 is shown in Figure 1.9.

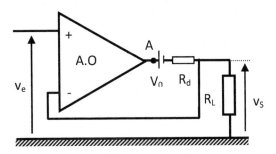

Figure 1.9. *Behavior of a "thresholdless diode" circuit when input voltage is positive*

The operational amplifier has very high differential impedance. The current going in or out of the two inputs e^+ and e^- (non-inverting and inverting input, respectively) is practically zero. Load resistance (R_L) is chosen very high compared to the dynamic resistance (R_d) of the diode. Then, the following relations can be written:

$$v_s = v_e = R_L i; \quad v_A = V_0 + (R_d + R_L)I; \quad v_A \cong V_0 + v_s$$

1.4.2. *Negative input voltage*

When voltage v_e is positive, output voltage follows input voltage. Before passing to the negative state, v_e necessarily passes through zero, and the same applies to output voltage v_s. Voltage $v_s = 0$ serves as reference for comparison with respect to $v_e < 0$. In this situation, the output of the operational amplifier passes to low saturation: $v_A = -V_{cc}$.

Diode D is blocked. The diagram of the "thresholdless" diode with operational amplifier is shown in Figure 1.10.

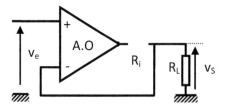

Figure 1.10. *State of the "thresholdless" diode when input voltage is negative*

There is no current across resistance R_L: $v_s = 0$

Various signals involved in the circuit of a "thresholdless" diode or ideal diode are presented in Figure 1.11.

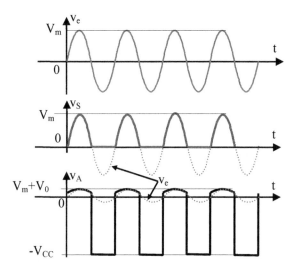

Figure 1.11. *Various signals involved in the "thresholdless diode" circuit.*
For a color version of this figure, see www.iste.co.uk/haraoubia/nonlinear1.zip

1.5. Load line and operating point

When a nonlinear component is inserted in an electronic circuit, it is important to know the voltage at its terminals and also the current across it. For example, when considering the circuit in Figure 1.12, which contains a diode (nonlinear element), the following equation can be written:

$$v_e = R.I_D + V_D$$

This equation with two unknowns should be solved in order to deduce I_D (current across the diode) and V_D (voltage at the diode terminals).

This ambiguity can be clarified using the relation that defines the variations of I_D as function of V_D for a diode:

$$I_D = I_{DS}(e^{\frac{nV_D}{V_T}} - 1)$$

Reaching a result requires several calculation stages. A simpler approach is possible however, and this involves a graphical solution to the problem. This approach uses the equation of the load line resulting from the circuit shown in Figure 1.12. At the level of the circuit mesh, the equation of the static load line is defined by the following relation:

$$I_D = \frac{v_e - V_D}{R}$$

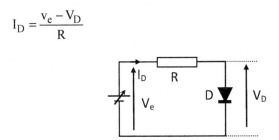

Figure 1.12. *Basic circuit for the definition of the load line equation*

If v_e and R are considered constant at a given instant, then the plot of I_D as a function of voltage V_D represents a line with negative slope, as shown in Figure 1.13. This is the load line. In order to draw this line, it is sufficient to know the coordinates of two points A and B as follows:

$$A \begin{pmatrix} I_D = \dfrac{v_e}{R_L} \\ V_D = 0 \end{pmatrix} \quad \text{and} \quad B \begin{pmatrix} I_D = 0 \\ V_D = v_e \end{pmatrix}$$

$$M \begin{pmatrix} I_{DM} \\ V_{DM} \end{pmatrix}$$

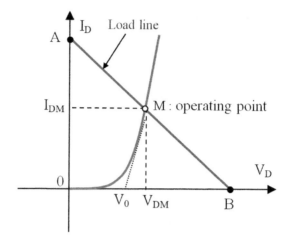

Figure 1.13. *Load line and operating point. For a color version of this figure, see www.iste.co.uk/haraoubia/nonlinear1.zip*

The intersection of this load line with the forward characteristic of the diode allows for the definition of the operating point of the diode (point M). Current I_{DM} is the current across the diode, and V_{DM} is the effective voltage at its terminals.

1.6. Other nonlinear components

1.6.1. *Thermistors or NTC (Negative Temperature Coefficient)*

A thermistor is a nonlinear component, the resistance of which varies as a function of temperature. When temperature increases, resistance decreases (Figure 1.14).

This type of component is used as a temperature sensor for various applications. Resistance variation as a function of temperature T approximately follows the following relation:

$$R = Ae^{\frac{B}{T}}$$

where A and B are two constants.

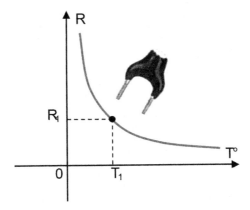

Figure 1.14. *Variation of the resistance of an NTC as a function of temperature*

1.6.2. *Photoresistors*

Photoresistors, also known as LDR (light-dependent resistors), are components made of semiconductors.

A photoresistor is sensitive to light. Its resistance decreases when lighting increases (Figure 1.15). Photoresistors have multiple uses, for example, automatic door opening.

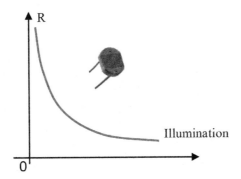

Figure 1.15. *Variation of the resistance of an LDR as a function of illumination*

A further application is for illumination control.

In order to illustrate the functionality of a photoresistor (LDR), the latter can be used for voltage control by light intensity, as indicated by the circuit shown in

Figure 1.16. The photoresistor (LDR) is inserted in a very simple circuit. The output voltage v_S varies as a function of the resistance of the photoresistor.

$$v_S = \frac{R}{R + R_{LDR}} E$$

$$R \ll R_{LDR}, v_S \cong 0$$

$$R \gg R_{LDR}, v_S \cong E$$

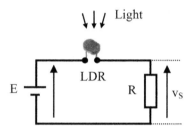

Figure 1.16. *Voltage v_S varies with LDR resistance*

1.6.3. *Varicap diodes or variable capacitance diodes*

The "varicap" diode is a junction diode. It is a device whose capacitance varies with reverse voltage "v_R" across its terminals.

The electric diagram representing a "varicap" diode is shown in Figure 1.17.

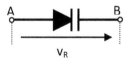

Figure 1.17. *Electric diagram of a variable capacitance diode*

The variation of capacitance as a function of voltage v_R is nonlinear:

$$C_j = \frac{C_{j0}}{\left(1 + \dfrac{v_R}{V_0}\right)^n}$$

where C_{j0} is the junction capacitance in the absence of external voltage; $V_o \cong 0.7$ V (for silicon) and n is a constant that depends on the diode fabrication process: n = 0.36 (diffusion diodes), n = 0.50 (planar diodes) and n > 0.50 (diodes made by special fabrication processes). Variable capacitance diodes are widely used in VCO (voltage-controlled oscillator). An example of the characteristics of MV1401 and MV1404 (Motorola) varicap diodes is shown in Figure 1.18.

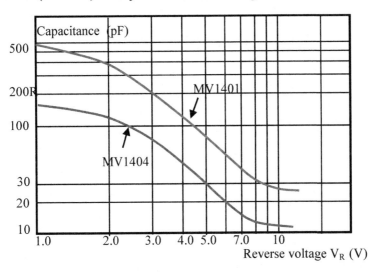

Figure 1.18. *C-V characteristic of Varicap diodes MV1401 & MV1404 (Motorola) - Source: Motorola Datasheet. For a color version of this figure, see www.iste.co.uk/haraoubia/nonlinear1.zip*

The variable capacitance diode MV1401 has a maximal capacitance (C_{max}) of about 600 pF and a minimal capacitance (C_{min}) of about 25 pF.

This results in a very significant ratio (above 20). It has a quality factor "Q" of about 200.

1.7. Nonlinear applications of the diode

1.7.1. *Half-wave rectification*

It should be recalled that the characteristic of the diode in the first approximation is very close to the characteristic of a real diode (at least for signals with high amplitude). Therefore, the study model to be considered here is that of the diode in the first approximation. It is known that the diode features a threshold voltage V_0, dynamic resistance R_d and reverse resistance R_i (very high resistance compared to the load resistance. The reverse resistance R_i can be assimilated to an open circuit).

A half-wave rectification circuit and the diode operation conditions are schematically presented in Figure 1.19.

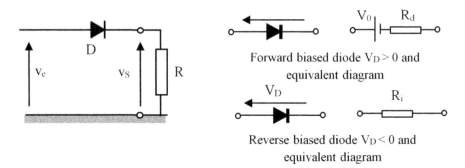

Forward biased diode $V_D > 0$ and equivalent diagram

Reverse biased diode $V_D < 0$ and equivalent diagram

Figure 1.19. *Half-wave rectification circuit and equivalent diagram of the diode depending on polarization*

1.7.1.1. *Forward-biased diode*

A forward-biased diode is equivalent to a voltage source V_0 in series with a resistance R_d. The detection circuit of a conducting diode is schematically presented in Figure 1.20.

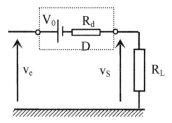

Figure 1.20. *Equivalent circuit of half-wave rectifier for a forward biased diode*

The definition of the output voltage v_S will be deduced from the equation governing voltages across the mesh. For the sake of simplicity, let us assume that the input voltage is a sine wave voltage of the form $v_e = V_M.\sin(\omega t + \varphi)$, where $\varphi = 0$.

According to the circuit shown in Figure 1.20, the following relations can be written:

$$v_e = V_0 + R_d \cdot i_D + v_s \quad \text{and} \quad v_s = R_L i_D$$

It should be recalled that the dynamic resistance R_d of the diode is very low, of the order of a few tens of ohms. Resistance R_L is chosen sufficiently high compared to R_d ($R_L \gg R_d$; $\gg$: much greater).

The expression of current i_D as a function of input voltage is:

$$i_D = \frac{v_e - V_0}{R_L + R_d} \cong \frac{v_e - V_0}{R_L}$$

Under these conditions, when the diode conducts, the output voltage is defined as:

$$v_S = v_e - V_0$$

The output voltage v_s practically exists only when $v_e > V_0$. As long as $0 < v_e < V_0$, $v_S = 0$. When $v_e = V_m$, $v_s = V_m - V_0$.

Thus, the peak value of the output signal equals the difference between the peak value of the input signal and the threshold voltage of the diode.

1.7.1.2. Reverse-biased diode

When the diode is reverse biased, it is equivalent to its reverse resistance. Under these conditions, the diagram of half-wave rectification is presented in Figure 1.21.

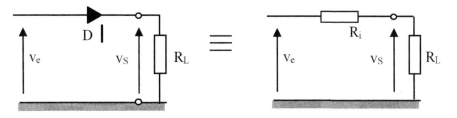

Figure 1.21. *Half-wave rectification circuit and its equivalent diagram for reverse biased diode*

The following is the expression of output voltage (reverse-biased diode):

$$v_S = \frac{v_e.R_L}{R_L + R_i}$$

Since R_i is very high, it can be considered that $R_i \gg R_L$: $v_S \cong 0$

A plot of the evolution of output voltage versus the evolution of the input voltage is presented in Figure 1.22.

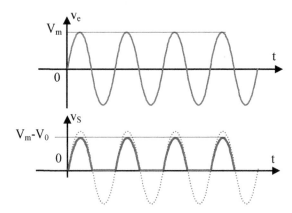

Figure 1.22. *Evolution of the output voltage for a half-wave rectification circuit. For a color version of this figure, see www.iste.co.uk/haraoubia/nonlinear1.zip*

1.7.2. Full-wave rectification with diode bridge

1.7.2.1. Principle

The full-wave rectification circuit (Figure 1.23) uses a diode bridge.

When a sine wave voltage is applied at the input of the rectification device, the resulting signal has its positive half-cycle preserved, but negative half cycle reversed.

For the sake of simplicity, let us consider two ideal diodes (threshold voltage and dynamic resistance are considered zero, while reverse resistance is assumed infinite).

1.7.2.2. *Operation*

Case of $v_e > 0$

Initially, the output voltage is zero. When the input voltage is positive, diodes D_1 and D_3 conduct, whereas diodes D_2 and D_4 are blocked.

Under these conditions, the equivalent diagram of a full-wave rectification circuit can be presented as shown in Figure 1.24.

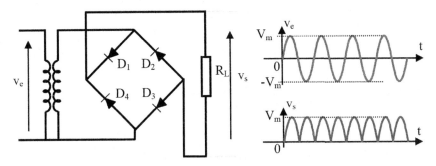

Figure 1.23. *Full-wave rectification circuit. For a color version of this figure, see www.iste.co.uk/haraoubia/nonlinear1.zip*

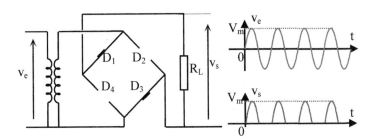

Figure 1.24. *Rendering of the positive half cycle at output. For a color version of this figure, see www.iste.co.uk/haraoubia/nonlinear1.zip*

Therefore, the positive half cycle is rendered at the output. It can be readily noted that when input voltage is positive:

$$v_s = v_e.$$

Case of $v_e < 0$

When submitted to the negative half-cycle of v_e, the diodes D_1 and D_3 are blocked and diodes D_2 and D_4 conduct, as shown in Figure 1.25. Under these conditions, the following relation can be written:

$$v_s = -v_e$$

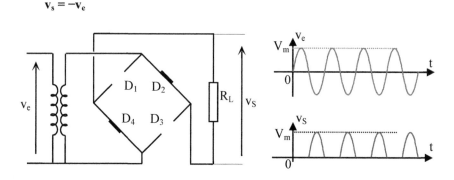

Figure 1.25. *Negative half cycle rectification. For a color version of this figure, see www.iste.co.uk/haraoubia/nonlinear1.zip*

IN SUMMARY.–

Since diodes D_1, D_2, D_3 and D_4 operate simultaneously, full-wave rectification is obtained at output (case of ideal diodes).

$$v_e > 0 \qquad v_s = v_e$$

$$v_e < 0 \qquad v_s = -v_e$$

$$v_s = |v_e|$$

NOTE.– For the sake of simplicity, and in order to facilitate understanding of circuit operation, this study of the full-wave rectification circuit does not mention the conduction threshold of the various diodes involved. Nevertheless, each rectified half-cycle is influenced by the conducting diodes. Thus, the amplitude of the half-cycle recovered at the output is reduced by twice the threshold voltage of a diode (the diode bridge is assumed to use four identical diodes).

1.7.2.3. *Full-wave rectification with thresholdless diodes*

To avoid the diode threshold problem and allow for full-wave rectification of low-amplitude signals, the circuit shown in Figure 1.26 can be used.

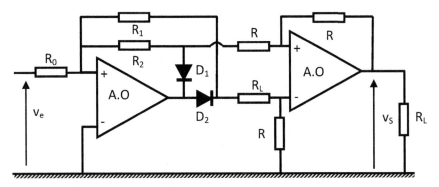

Figure 1.26. *Full-wave rectification circuit with thresholdless diodes*

The circuit is composed of two thresholdless detectors similar to those studied in section 1.4 (design of thresholdless diode).

One of the detectors recovers the positive half-cycle and the second the negative half-cycle. The detection circuit is followed by a subtractor assembly, the role of which is to reverse the polarity of the negative half-cycle.

The signals involved in the circuit are presented in Figure 1.27.

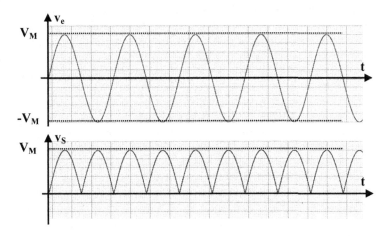

Figure 1.27. *Signals involved in the full-wave rectification circuit*

NOTES.– Unlike the rectification circuit with diode bridge, the device presented in Figure 1.26 does not require a floating ground and avoids the diode's threshold-related problem.

Proper functioning of circuit 1.26 requires $R_1 = R_2$.

Resistance R_1 must be very low compared with resistance R. Otherwise, the rectified half-cycles will not have the same amplitude simply because resistance R_1 is loaded by a resistance equal to 2R.

The response curve of the full-wave rectification circuit shown in Figure 1.26 is presented in Figure 1.28.

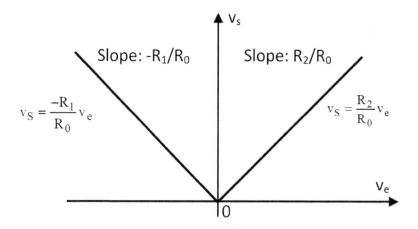

Figure 1.28. *Response curve of the full-wave rectification circuit with thresholdless diodes*

Full-wave rectification obviously requires $R_1 = R_2 = R_0$. This yields at the output: $\mathbf{v_S} = |\mathbf{v_e}|$

In order to avoid the problem of dynamics of resistances and to have a perfect full-wave rectification, the diagram shown in Figure 1.26 can be modified to a certain extent, to result in Figure 1.29. All resistances have been chosen equal. The operation of the full-wave detection circuit (or of the absolute value of the input voltage) will be explained in the following. For the sake of simplicity, the signal v_e applied at the detection circuit input is assumed to be a sine wave signal.

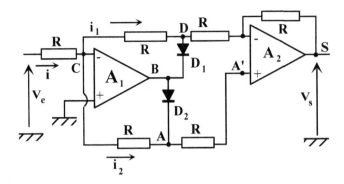

Figure 1.29. *Circuit for detecting the absolute value of a signal*

When the voltage applied at input is positive, diode D_1 is conductive and diode D_2 is equivalent to an open circuit. Therefore, the diagram of the detection circuit is as shown in Figure 1.30.

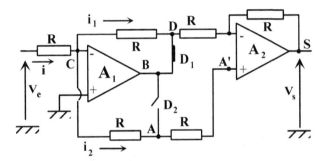

Figure 1.30. *State of the circuit for the detection of the absolute value for $v_e > 0$*

$v_e > 0$

An analysis of the circuit shown in Figure 1.30 reveals that voltage at point C is zero. The same is valid for voltage at point A.

The following relations can be written: $v_A = 0$; $v_s = -v_B$ and $v_B = -v_e$

Finally, $v_s = v_e$

When input voltage is negative ($v_e < 0$), diode D_2 conducts and diode D_1 is blocked. Therefore, the circuit for the detection of the absolute value is reduced to the circuit schematically shown in Figure 1.31.

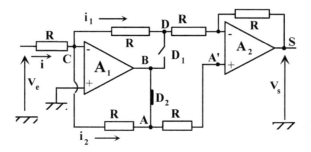

Figure 1.31. *State of the circuit for the detection of the absolute value when $v_e < 0$*

$v_e < 0$

According to the diagram in Figure 1.31, it can be noted that voltage at point C is zero. Similarly, the following can be written:

$$v_A = -R.i_2; \; v_{A'} = v_A; \; v_D = \frac{v_A}{2}; \; v_e = R(i_1 + i_2); \; i = (i_1 + i_2);$$

$$v_e = -\frac{3}{2}v_A;$$

$$v_s = -Ri1 + v_A;$$

$$v_s = \frac{3}{2}v_A.$$

Finally, $\mathbf{v_s = -v_e}$.

IN SUMMARY.–

The following can be written:

$$\left.\begin{array}{ll} v_e < 0 & v_s = -v_e \\ v_e > 0 & v_s = v_e \end{array}\right\} \Rightarrow \quad v_s = |v_e|$$

When a sine wave signal v_e (see Figure 1.32(a)) is applied at input, the output signal has the form schematically presented in Figure 1.32(b).

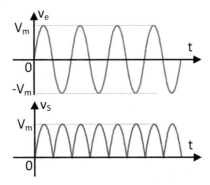

Figure 1.32. *a) sine wave input signal; b) signal at the output of the detector of absolute value. For a color version of this figure, see www.iste.co.uk/haraoubia/nonlinear1.zip*

1.7.3. Peak clipping

If one part of the signal has to be eliminated (nonlinear transformation), for example, when transforming a sine wave signal into a signal that closely resembles a square wave signal, the diode is a component that can meet this specific need. The levels of peak clipping can therefore be adjusted as shown in Figure 1.33. Proper operation of the peak clipping requires the amplitude of the signal to be clipped to be above the amplitudes V_1 and V_2. In this approach, the conduction threshold of diodes D_1 and D_2 has been neglected.

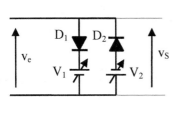
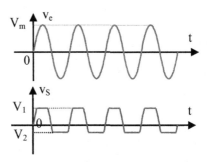

Figure 1.33. *Principle of signal peak clipping. For a color version of this figure, see www.iste.co.uk/haraoubia/nonlinear1.zip*

1.7.4. *Peak detector*

To be able to detect the maximal value of an alternating signal, the circuit shown in Figure 1.34 can meet this need in a satisfactory manner. For an easier understanding of the operation of this device, let us suppose that the diode is ideal and the input signal is a sine wave: $v_e = V_m.\sin(\omega t)$.

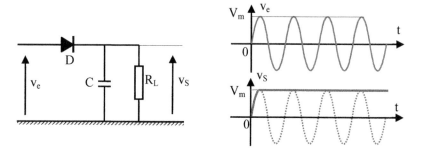

Figure 1.34. *Principle of peak detector and signals involved. For a color version of this figure, see www.iste.co.uk/haraoubia/nonlinear1.zip*

During the positive half-cycle, the diode is in conduction state. Capacitor C charges very rapidly through the dynamic resistance of the diode at peak value V_m.

The result at the output is $v_S = V_m$

In the presence of the negative half-cycle, the diode is blocked and the capacitor tries to discharge through R_L. The value of this resistance is chosen high in order to avoid the capacitor discharging during this negative half-cycle.

Thus, the result at output is a continuous voltage, the amplitude of which is equal to that of the peak value of the input signal. The arrival of the second positive half-cycle allows the capacitor to resume charging and thus compensate for the possible partial ensuing discharge.

When the detection of the peak value of a variable signal of constant amplitude is intended, the discharging time constant ($\tau_d = R_L C$) should be far greater than the period of the input signal.

However, if this time constant is too high and the signal amplitude at input decreases, then there is a risk that the circuit does not follow this change and detection is erroneous.

A too low time constant will not allow for the detection of signal peak at input. There will be ripples connected to the discharging and charging of capacitor C and across the load resistance R_L and the dynamic resistance of the diode.

Therefore, a compromise is needed with respect to the discharging constant of the capacitor. The conduction threshold of the diode should also be taken into account in practice.

As indicated, and in order to facilitate simplification and understanding, the diode has been assumed ideal. In reality, a continuous voltage v_S slightly lower than V_m is obtained:

$v_S = V_m - V_0$, where V_0 is the threshold voltage of the diode.

1.7.5. Recovery circuits

1.7.5.1. Principle

A recovery circuit allows a continuous component to be added to a variable signal, as shown in Figure 1.35.

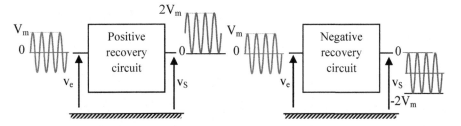

Figure 1.35. Principle of the recovery of a continuous component. For a color version of this figure, see www.iste.co.uk/haraoubia/nonlinear1.zip

The continuous component can be positive or negative.

The example of how one of the two circuits in Figure 1.36 is used allows a positive or negative continuous component to be rendered depending on the diode connection direction.

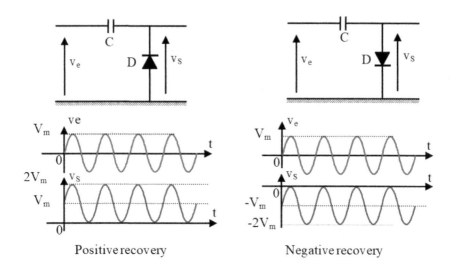

Positive recovery   Negative recovery

Figure 1.36. *Circuits for the recovery of the continuous component of diodes.*
For a color version of this figure, see www.iste.co.uk/haraoubia/nonlinear1.zip

For a simpler explanation of the operation of these circuits, let us assume that the input signal is a sine wave and its average value is zero. The circuits for the recovery of the positive component and the negative component have the same operating principle.

1.7.5.2. Operation

Because of the similarity of their operation, only one of the two circuits will be studied. Let us consider the example of the circuit that allows for the recovery of the negative continuous component.

For this purpose, the diode will be assumed ideal. The input signal is a sine wave: $v_e = V_m \sin(\omega t)$

When voltage v_e is positive, the diode is in conduction state. For the sake of simplicity, it can be assimilated to a short circuit (ideal diode), as indicated in the circuit shown in Figure 1.37.

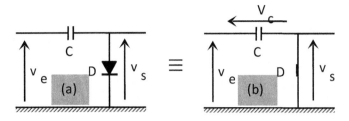

Figure 1.37. *Circuits for the recovery of the negative continuous component and equivalent diagram when the diode conducts*

$V_e > 0$

Under these conditions, the following relations can be written:

$$v_S = 0; \; v_e = V_C + v_S = V_C; \; v_S = v_e - V_C.$$

Capacitor C charges to the peak value V_m ($V_c = +V_m$). When voltage v_e drops slightly below the peak value V_m, the voltage between the anode and the cathode of the diode is negative and, consequently, the diode is blocked. The equivalent recovery circuit at this moment is schematically presented in Figure 1.38. It can then be written that:

$$v_S = v_e - V_C; \; V_C = V_m; \; v_S = v_e - V_m$$

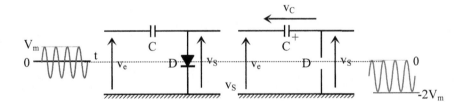

Figure 1.38. *The equivalent recovery circuit when the capacitor charges and output signal. For a color version of this figure, see www.iste.co.uk/haraoubia/nonlinear1.zip*

It can thus be noted that the output signal is none other than the input signal, to which a continuous negative voltage of amplitude equal to "$-V_m$" is added.

1.7.6. *Influence of dynamic and reverse resistances on the recovery of the continuous component of a signal*

1.7.6.1. *General context*

In this context, let us suppose that the diode introduces a direct resistance R_d and a reverse resistance R_i. The excitation source has an output resistance R_g (internal resistance). The diode operation characteristic can be schematically presented as shown in Figure 1.39.

For the sake of simplicity, the threshold voltage V_0 of the diode is assumed zero. The recovery circuit is represented in Figure 1.40. Load resistance R_L is chosen so that:

$$R_i \gg R_L \gg R_d$$

The diode can be alternatively conducting or blocked depending on the state of the input signal, and the diagrams for each case (conducting or blocked diode) are given in Figures 1.41(a) and 1.41(b), respectively.

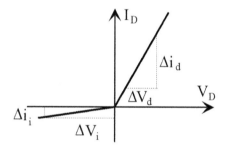

Figure 1.39. *Characteristic of the diode. Threshold voltage V_0 is assumed zero nulle*

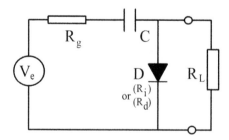

Figure 1.40. *Recovery circuit*

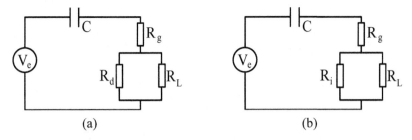

Figure 1.41. *a) Equivalent diagram of the recovery circuit when the diode conducts; b) equivalent diagram of the recovery circuit when the diode is blocked*

Dynamic and reverse resistances are defined by the following relations:

$$R_d = \frac{\Delta V_d}{\Delta i_d}; \quad R_i = \frac{\Delta V_i}{\Delta i_i}$$

The circuits presented in Figure 1.41 are first-order systems.

Time constants (τ_1 and τ_2) are defined as follows:

– Conducting diode: $\tau_1 \cong (R_g + R_d).C$

Resistance R_L is assumed very high compared to R_d.

– Blocked diode: $\tau_2 \cong (R_g + R_L).C$

Resistance R_L is assumed low compared to reverse resistance R_i. The input signal is assumed rectangular with non-zero average value (Figure 1.42).

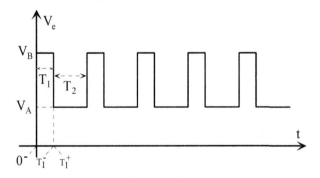

Figure 1.42. *Signal applied at recovery circuit input*

The capacitor tends to get charged during the high state of the input signal and discharges during the low state. It should be noted that the studied circuit allows for the recovery of the negative continuous component. Therefore, it can be assumed that the form of the output signal is almost similar to that schematically shown in Figure 1.43.

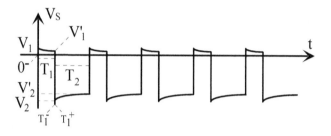

Figure 1.43. *Shape of the output signal*

The values or expressions of voltages V_1, V'_1, V_2 and V'_2 will be provided with precision upon the following calculation.

In order to find the expression and the exact shape of the output signal, the effect of the source resistance R_g must also be taken into account. It is worth recalling that a choice has been made to consider:

$$R_i \gg R_L \gg R_d$$

The diagrams used for determining the exact shape of the output signal are shown in Figures 1.44(a) and 1.44(b).

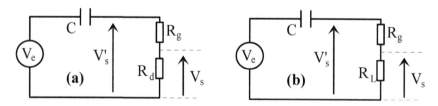

Figure 1.44. *State of the recovery circuit for a) conducting diode; b) blocked diode if $R_i \gg R_L \gg R_d$*

At t = 0⁻, v_e(t) = V_A (see Figure 1.42), and the diode is blocked. The expression of the output signal is given by $v_s(0^-) = V'_2$.

Voltage V_{c1} across the capacitor is defined as:

$$v_e - v'_s = v_{c1} = v_e - v_s \frac{R_g + R_L}{R_L}$$

$$v_{c1} = V_A - V'_2 \frac{R_g + R_L}{R_L}$$

At t = 0⁺, the amplitude of the input voltage is defined as V_B. At this moment, the diode becomes conducting:

$$v_s(t) = V_1$$

The value of the output voltage presents an abrupt shift: $V'_2 \rightarrow V_1$.

The new voltage v_{c2} across the capacitor is now defined by the following relation:

$$v_e - v'_s = v_{c2} = v_e - v_s \frac{R_g + R_d}{R_d}$$

$$v_{c2} = V_B - V_1 \frac{R_g + R_d}{R_d}$$

It should be noted that the voltage across the capacitor can in no way vary instantaneously between instants t = 0⁻ and t = 0⁺. Therefore, the following can be written:

$$v_{c1} = v_{c2}$$

$$V_A - V'_2 \frac{R_g + R_L}{R_L} = V_B - V_1 \frac{R_g + R_d}{R_d}$$

$$V_B - V_A = V_1 \frac{R_g + R_d}{R_d} - V'_2 \frac{R_g + R_L}{R_L}$$

At $t = T_1$, the same calculation as the previous one can be performed. Thus, at instant $t = T_1^-$, the diode is conducting: $v_e = V_B$; $v_s = V'_1$

$$v'_{c1} = V_B - V'_1 \frac{R_g + R_d}{R_d}$$

where v'_{c1} is the voltage across the capacitor at instant $t = T_1^-$. On the contrary, at t $= T_1^+$, the diode is blocked: $v_e = V_A$; $v_s = V_2$

$$v'_{c2} = V_A - V_2 \frac{R_g + R_L}{R_L}$$

As has already been noted, the voltage across the capacitor cannot vary instantaneously: $v'_{c1} = v'_{c2}$

$$V_B - V_A = V'_1 \frac{R_g + R_d}{R_d} - V_2 \frac{R_g + R_L}{R_L}$$

In the time period ($0^+ < t < T_1^-$), the diode is conducting and the capacitor charges with a time constant equal to $(R_g + R_d).C$. The variation in output voltage evolves according to an exponential law, the value of which at $t = T_1$ is:

$$v_s(T_1) = V'_1 = V_1.e^{\frac{-T_1}{(R_g + R_d)C}}$$

In the interval $T_1 < t < T_1 + T_2$, the diode is blocked. The capacitor discharges with a time constant equal to $(R_g + R_L)C$. At T_2, the output voltage has the value $v_s(T_2) = V'_2$:

$$v_s(T_2) = V'_2 = V_2.e^{\frac{-T_2}{(R_g + R_L)C}}$$

It can thus be noted that using only the amplitudes of the input signal (V_A and V_B) and the value of resistances R_g, R_d and R_L, the various amplitudes of the output signal, namely V_1, V'_1, V_2 and V'_2, can be determined accurately. The relationship between V'_1 and V_1 as well as that between V'_2 and V_2 provide the exact shape of the output signal.

1.7.6.2. Specific cases

Let us assume that the source resistance (or internal resistance of the generator) has no influence: $R_g = 0$. Under these conditions:

$$V_B - V_A = V = V_1 - V'_2 = V'_1 - V_2.$$

The amplitudes of the output signal discontinuities are equal to the amplitudes of the input signal discontinuities.

On the contrary, when $R_g \neq 0$, it can be noted that the amplitudes of the output signal discontinuities ($V_1 - V'_2$) and ($V'_1 - V_2$) are smaller than the amplitudes of the input signal discontinuities ($V_B - V_A$). Attenuation has therefore taken place.

Let us consider:

$$\Delta V_d = V_1 - V'_1; \qquad \Delta V_i = V'_2 - V_2$$

where ΔV_d is the amplitude of the discontinuity when the diode is conducting and ΔV_i is the amplitude of the discontinuity when the diode is blocked.

It can be noted that:

$$[V_1 - V'_1]\frac{R_d + R_g}{R_d} = [V'_2 - V_2]\frac{R_L + R_g}{R_L}$$

$$\Delta V_d = \Delta V_i \left[\frac{R_L + R_g}{R_L} \frac{R_d}{R_d + R_g} \right]$$

In practice:

$$R_L \gg R_g: \Delta V_d = \Delta V_i \left[\frac{R_d}{R_d + R_g} \right]$$

when $R_d \gg R_g$, $\Delta V_d = \Delta V_i$

1.7.6.3. *Necessary conditions for perfect recovery*

For perfect recovery of a continuous component, a capacitor C with high capacitance must be employed.

This is a necessary condition for avoiding the capacitor C having the time to discharge when the diode is in blocked state.

On the contrary, the capacitor should be able to charge instantaneously (that imposes $R_g = R_d = 0$) during buildup time (diode in conducting state).

The conditions set forth cannot be simultaneously achieved. Thus, in order to obtain an acceptable form of output signal, it is important to adopt a compromise for the charging and discharging time constants of the capacitor C.

$\tau_1 = (R_g + R_d)C \ll T_1$, where τ_1 is the charging time constant

$\tau_2 = (R_g + R_L)C \gg T_2$, where τ_2 is the discharging time constant.

Durations T_1 and T_2 are two time intervals imposed by the shape of the signal applied at the input of the recovery circuit (Figure 1.45).

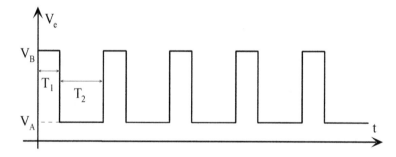

Figure 1.45. *Input signal and durations of the high state T_1 and low state T_2*

When this signal is applied at the input of the studied recovery circuit, and depending on the conditions imposed on the charging and discharging time constants, the shape of the output signal can be approximated as schematically represented in Figure 1.46.

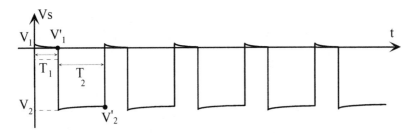

Figure 1.46. *Output signal of the recovery circuit when:* $\tau_1 \ll T_1$ *and* $\tau_2 \gg T_2$

The capacitor charges very rapidly ($\tau_1 \ll T_1$), which leads to $V'_1 = 0$. The previously established relations are simplified: $V'_1 = 0$ and $\Delta V_d = V_1$:

$$V_B - V_A = V'_1 \frac{R_g + R_d}{R_d} - V_2 \frac{R_g + R_L}{R_L}$$

$$V_2 = -(V_B - V_A)\frac{R_L}{R_g + R_L}$$

Furthermore, it is known that:

$$v_s(T_2) = V'_2 = V_2 . e^{\frac{-T_2}{(R_g + R_L)C}}$$

Time constant τ_2 is very big compared to the small state duration.

$$\tau_2 = (R_g + R_L).C \gg T_2$$

$$v_s(T_2) = V'_2 = V_2 . \left[1 - \frac{T_2}{(R_g + R_L)C} \right] = V_2 . \left[1 - \frac{T_2}{\tau_2} \right]$$

The amplitude of the output voltage variation when the diode is blocked is defined as $\Delta V_i = V'_2 - V_2$. The positive amplitude of the output signal can be expressed as a function of ΔV_i.

$$V_1 = \Delta V_d = \Delta V_i \left[\frac{R_L + R_g}{R_L} \frac{R_d}{R_d + R_g} \right] ; \quad V_1 = \frac{R_d(R_L + R_g)}{R_L(R_d + R_g)}(V'_2 - V_2)$$

Finally, the amplitude of the positive part of the output signal is related to the amplitude of the input signal discontinuity, the discharging time constant and the duration of the low state of the input signal:

$$V_1 = (V_B - V_A) \frac{R_d}{(R_d + R_g)} \frac{T_2}{\tau_2}$$

When $T_2 \ll \tau_2$, perfect recovery will ensue, since $V_1 \to 0$.

NOTE.– Even when a capacitor of very high capacity is chosen, it is impossible to obtain $R_g = R_d = 0$.

When the diode is conducting, the output signal is expressed as the sum of a fraction of the input signal and a constant:

$$v_s(t) = v_e(t) \frac{R_d}{R_d + R_g} + \text{cst.}$$

1.7.7. *Logarithmic amplifier*

The diode is a nonlinear component. The current–voltage characteristic is expressed as $I_D = I_{DS}(e^{\frac{V_D}{V_T}} - 1)$

where I_D is the current across the diode, V_D is the voltage across the diode, I_{DS} is the reverse current and $V_T = 26$ mV at 300 K.

When operating at high values, the expression of the current I_D across the diode as a function of voltage V_D across the diode can be reduced to $I_D = I_{DS}.e^{\frac{V_D}{V_T}}$

In order to realize a logarithmic amplifier, a circuit with operational amplifier and diode shown in Figure 1.47(a) is used.

Point M' is a virtual ground (specific to operational amplifiers). When the input voltage is positive, the expressions of input voltage and output voltage are given by:

$$v_e = RI_D \quad \text{and} \quad v_S = v_D$$

$$v_D = V_T.\text{Ln}(\frac{I_D}{I_{DS}} - 1)$$

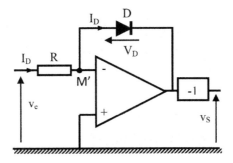

Figure 1.47(a). *Logarithmic amplifier with diode*

It is worth noting that large values are involved. The direct current I_D is very high compared to the reverse current I_{DS}:

$$v_D \cong V_T.Ln(\frac{I_D}{I_{DS}}) = V_T.Ln(\frac{v_e}{RI_{DS}}) \quad \text{and} \quad v_S \cong -V_T.Ln(\frac{v_e}{RI_{DS}})$$

It is quite obvious that this circuit cannot operate as logarithmic amplifier unless v_e is positive. When v_e is negative, the direction of the diode should definitely be reversed if the circuit is to operate as logarithmic amplifier.

Figure 1.47(b) shows an example of signals employed in a logarithmic amplifier.

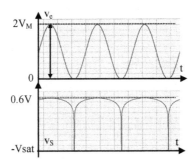

Figure 1.47(b). *Signals employed in a logarithmic amplifier with diode*

1.7.8. Anti-logarithmic amplifier

In order to realize an anti-logarithmic amplifier, it suffices to reverse the roles of resistance and diode in the logarithmic amplifier. The connection of the diode depends on the sign of the input signal to be processed. For example, when the signal of "v_e" is positive, the diagram in Figure 1.48 is used.

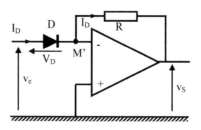

Figure 1.48. *Anti-logarithmic amplifier with diode $v_e > 0$*

Point M' is a virtual ground. It is for this reason that the input voltage should be positive for the circuit to operate as an antilog amplifier.

Otherwise, the output will be zero, as the diode remains blocked.

$v_e > 0$:

$$v_S = -RI_D \quad \text{and} \quad v_e = V_D$$

$$v_S = -RI_{DS}(e^{\frac{Ve}{V_T}} - 1) \cong -RI_{DS}.e^{\frac{Ve}{V_T}}$$

If the input voltage "v_e" is negative, the connection direction of the diode will be reversed, as shown in the diagram in Figure 1.49.

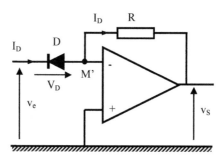

Figure 1.49. *Anti-logarithmic amplifier with diode when $v_e < 0$*

The following relations can be written as:

$$v_S = RI_D \quad \text{and} \quad v_e = -V_D;$$

$$v_S = RI_{DS}(e^{\frac{-Ve}{V_T}} - 1) \cong RI_{DS}.e^{\frac{-Ve}{V_T}}$$

1.7.9. *Logic functions with diodes*

Logic functions can be readily built using exclusively diodes. The focus here will be on simple functions, such as "OR" or "AND".

1.7.9.1. *Principle of a logic function OR*

The logic function "OR" is represented by the electric diagram and its truth table is shown in Figure 1.50.

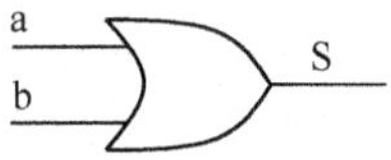

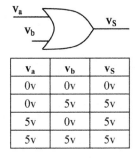

a	b	S
0	0	0
0	1	1
1	0	1
1	1	1

Figure 1.50. *Representation of the logic 'OR' function and its truth table*

The representation from a potential perspective is shown in Figure 1.51 (TTL circuits).

V_a	V_b	V_S
0v	0v	0v
0v	5v	5v
5v	0v	5v
5v	5v	5v

Figure 1.51. *Representation of the logic 'OR' function and its truth table: voltage perspective*

1.7.9.2. *Logic "OR" function with diodes*

The circuit shown in Figure 1.52 plays exactly the role of a logic "OR" function.

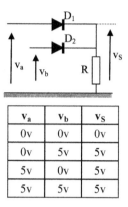

v_a	v_b	v_S
0v	0v	0v
0v	5v	5v
5v	0v	5v
5v	5v	5v

Figure 1.52. *Logic 'OR' gate with diodes*

In fact, it can be noted that when $v_a = 0v$ and $v_b = 0v$, the two diodes D_1 and D_2 are blocked and there is no current across R: $v_S = 0v$.

When $v_a = 5v$ and $v_b = 0v$, diode D_1 conducts and diode D_2 is blocked:

$v_S = v_a = 5v$.

When $v_a = 0v$ and $v_b = 5v$, diode D_1 is blocked and diode D_2 conducts:

$v_S = v_b = 5v$.

When $v_a = 5v$ and $v_b = 5v$, the two diodes D_1 and D_2 conduct and $v_S = 5v$. Furthermore, a summary of the above can be drawn, evidencing that the circuit with diodes is quite representative of a logic "OR" function.

1.7.9.3. *Principle of a logic AND function*

The representation of the electric diagram of an AND gate and its truth table are shown in Figure 1.53.

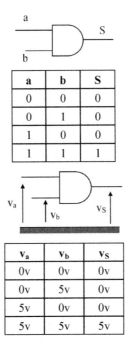

a	b	S
0	0	0
0	1	0
1	0	0
1	1	1

v_a	v_b	v_S
0v	0v	0v
0v	5v	0v
5v	0v	0v
5v	5v	5v

Figure 1.53. *Representation of the logic 'AND' function and its truth table*

1.7.9.4. "AND" gate with diodes

The circuit to be studied is shown in Figure 1.54. To facilitate understanding of the circuit operation, the diodes employed are considered ideal.

When $v_a = v_b = 0$, diodes D_1 and D_2 are conducting. Then, $\mathbf{v_S = v_a = v_b = 0}$ (Figure 1.54).

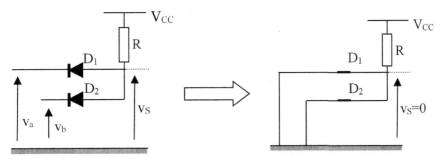

Figure 1.54. *Left: logic 'AND' gate with diodes.*
Right: state of the circuit when $v_a = v_b = 0$

When $v_a = 0$ and $v_b = V_{CC}$ ($v_a = V_{CC}$ and $v_b = 0$), diode D_1 is conducting and D_2 is blocked (diode D_1 is blocked and D_2 is conducting). Then, $v_S = 0$.

When $v_a = V_{CC}$, $v_b = V_{CC}$, diodes D_1 and D_2 are blocked (as shown in Figure 1.55), and there is no current across resistance R. Then, $v_S = V_{CC}$.

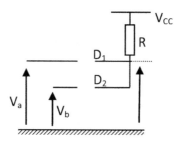

Figure 1.55. *State of the circuit when $V_a=V_b=V_{cc}$*

Summarizing the operation of the circuit shown in Figure 1.54 and considering all the states related to voltages v_a and v_b, the voltage v_s can be defined, as shown in Table 1.1.

Table 1.1 is quite representative of the function realized by a logic "AND" gate.

V_a	V_b	V_S
0v	0v	0v
0v	5v	0v
5v	0v	0v
5v	5v	5v

Table 1.1. *Table related to the state of the circuit shown in Figure 1.54*

1.8. Exercises

EXERCISE 1

The diode circuit shown in Figure E1.1 is the object of our study. As a first step, the diode is represented by its real model. The following are given: $I_S = 10^{-12}$ A (reverse saturation current); (KT/e = 26 mV); "**K**", the Boltzmann constant, "**T**", the temperature and "**e**", the elementary charge. The expectation is to have a current $I_1 = 1$ mA across the circuit from a voltage $v_e = 2$ V.

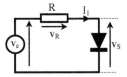

Figure E1.1.

1) Calculate the voltage across the diode when it is represented by its exponential model.

2) Determine the expression and value of R that would impose the current I_1.

3) Calculate the dynamic resistance R_d of the diode.

4) The diode is represented by a simplified model ($V_D = V_0 = 0.6$ V, $R_d = 0$ and R_i is infinite), where V_0 is the threshold voltage, R_d is the forward dynamic resistance and R_i is the reverse resistance. Calculate the current I_2 that flows through the mesh (the value of resistance R calculated at point 2 will be used). Compare the currents I_2 and I_1. Draw the conclusions.

5) Using the diode model described at point 4, $v_e = 4\sin(2\pi ft+\varphi)$, represent v_e, v_R and v_S.

EXERCISE 2

The circuit is given in Figure E2.1.

1) Write the equation of the load line.

2) Plot the direct characteristic of the diode together with the load line and evidence the operating point in the following cases:

2.1) Real diode; 2.2) diode in second approximation ($V_0 = 0.6$ V; $R_d = 0$ and R_i is infinite).

3) The diode is considered ideal ($V_0 = 0$; $R_d = 0$ and R_i is infinite). The studied circuit is shown in Figure E2.1. The voltage v_e is sinusoidal ($v_e = V_m.\sin(2\pi ft)$).

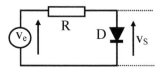

Figure E2.1.

3.1) Find the expression of $v_s(t)$ and represent $v_s(t)$.

3.2) Represent $v_s = f(v_e)$.

4) Let us consider that the diode is represented by its model in Figure E2.2. Answer the same questions as those in question 3.

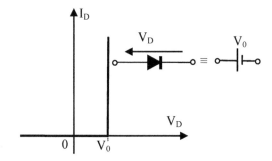

Figure E2.2.

5) Let us now consider the circuit shown in Figure E2.3. The diode is assumed ideal.

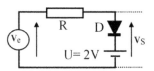

Figure E2.3.

5.1) Fill in Table E2.1.

v_e	State of the diode	v_s
0		
1		
1.5		
3		
4		
6		

Table E2.1.

5.2) Plot $v_s = f(v_e)$.

EXERCISE 3

Let us consider the circuit shown in Figure E3.1. Diodes D_1 and D_2 are identical.

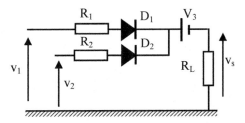

Figure E3.1.

Voltages v_1 and v_2 are continuous. The following are given: $v_1 = 12$ V, $v_2 = 8$ V and $V_3 = 5$ V. The threshold voltage of the diodes is $V_0 = 0.6$ V and $R_1 = R_2 = R_L = 10$ kΩ.

1) Determine the operating state of the two diodes and find the expression of the current through load resistance R_L and its numerical value. Determine the expression of v_s and its value.

2) Now, voltages v_1 and v_2 are sinusoidal. They have the same frequency and their initial phase is zero. Their amplitudes are different.

$v_1 = 12\sin(2\pi ft)$ and $v_2 = 8\sin(2\pi ft)$ and $V_3 = 5$ V. Find the expression of the output voltage and its representation as a function of time.

EXERCISE 4

Let us consider the circuit shown in Figure E4.1. The diode has the following characteristics: threshold voltage $V_0 = 0.6$ V, forward dynamic resistance $R_d = 0$ and infinite reverse resistance.

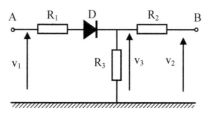

Figure E4.1.

1) As a first step, let us have $v_1 = 1$ V, $v_2 = 4$ V and $R_2 = R_3 = 1$ kΩ. Find the expression of voltage v_3 as a function of the circuit elements and the value of the current across resistance R_3.

2) The following are given: $v_1 = 3$ V, $v_2 = 3$ V and $R_1 = R_2 = R_3 = 1$ kΩ. The same question formulated at point 1 should be answered.

3) Let us now set $R_2 = 5$ kΩ and $R_1 = R_3 = 1$ kΩ. Points A and B are joined and the periodic signal s(t) is applied (Figure E4.2). Draw the graphic representation of signal $v_3(t)$.

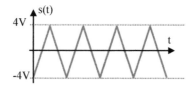

Figure E4.2.

EXERCISE 5

An ideal diode D ($V_0 = 0$, $R_d = 0$ and R_i is infinite) is inserted in the circuit shown in Figure E5.1.

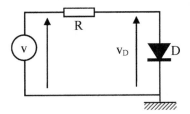

Figure E5.1.

1) Draw the representation of its characteristic $I_D = f(V_D)$.

2) Assuming that voltage $v = 12$ V and $R = 1$ kΩ, find the expression and the value of the maximum current flowing through diode D.

3) Write the equation of the load line and represent it on the same diagram as the characteristic $I_D = f(V_D)$.

4) Let us now consider the diagram in Figure E5.2. The task is to fill in Table E5.1, knowing that diodes D_1 and D_2 are considered ideal (for the diodes: evidence B for blocked state and P for conducting state).

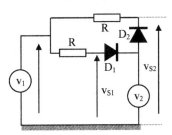

Figure E5.2.

v_1	v_2	State D_1	State D_2	v_{S1}	v_{S2}
2 V	5 V				
0 V	2 V				
−4 V	−2 V				
4 V	2 V				
−1 V	−3 V				
4 V	0 V				

Table E5.1.

5) Signal v_1 is sinusoidal with peak amplitude 3 V; $v_2 = 2$ V. Represent the evolutions of v_{S1} and v_{S2} as functions of time.

EXERCISE 6

In the circuit shown in Figure E6.1, D_1 and D_2 are assumed ideal ($V_0 = 0$, $R_d = 0$ and R_i is infinite) and $v_1 = 5\sin(2\pi ft)$.

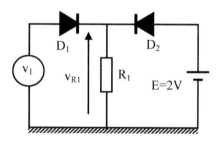

Figure E6.1.

1) Find the expression of voltage across R_1.

2) Represent the evolution in time of the voltage across resistance R_1.

EXERCISE 7

Let us consider a forward-polarized PN silicon junction. The direct characteristic of this junction follows the equation: $I_D = I_{S0} \left[e^{\frac{V_D}{V_{T0}}} - 1 \right]$; $V_{T0} = 26$ mV at T = 300 K.

At $T_1 = 350$ K, the following values are noted: $V_D = 0.3$ V, $I_D = 20$ mA.

1) Calculate V_{T1}.

2) Calculate the reverse saturation current I_{S1}. Compare I_{S1} and I_D and draw the conclusions.

3) Knowing that $I_S = I_{S0} \left[e^{a(T-T_0)} \right]$, where a = 0.06 for the semiconductor employed, calculate the reverse saturation current I_{S0} for a temperature $T_0 = 300$ K. Compare I_{S1} and I_{S0}.

4) Calculate the temperature variation ΔT in °C so that the reverse saturation current I_S doubles its value.

EXERCISE 8

The circuit schematically represented in Figure E8.1 is given in which $R_1 = R_2 = 100\ \Omega$.

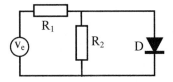

Figure E8.1.

1) Represent on the diagram voltages V_{R1}, V_{R2} and V_D (across R_1, R_2 and D, respectively), as well as currents I_1, I_2, I_D (across R_1, R_2 and D, respectively).

2) Find the value of v_e starting from which D is conducting.

3) Resistance R_2 is removed (Figure E8.2). Under these conditions, determine the equation of the load line, its graphic representation and the coordinates of the operating point for $v_{e1} = 1.2$ V.

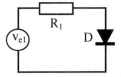

Figure E8.2.

4) v_{e1} is a triangular voltage (Figure E8.3). D is considered ideal ($V_0 = 0$, $R_d = 0$ and R_i is infinite) in this case. Represent one below the other the evolutions of v_{e1}, v_{R1} and v_D.

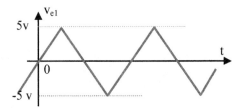

Figure E8.3.

EXERCISE 9

Let us consider the circuit shown in Figure E9.1.

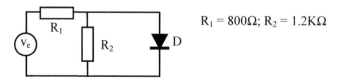

$R_1 = 800\Omega$; $R_2 = 1.2K\Omega$

Figure E9.1.

1) Initially, v_e is a continuous voltage, the diode is considered ideal (threshold voltage V_0 is zero, forward dynamic resistance R_d is zero and reverse resistance R_i is infinite). Calculate the current I across the diode for the following cases: $v_e = 2$ V, $v_e = 4$ V, $v_e = -0.5$ V, $v_e = -2$ V and $v_e = -4$ V. Then, deduce the curve $I = f(v_e)$.

2) Now, the diode features a conduction threshold $V_0 = 0.6$ V, resistance $R_d = 0$ and R_i is infinite. Calculate the current I across the diode for the cases: $v_e = 4$ V, $v_e = 2$ V, $v_e = 1$ V, $v_e = -0.5$ V, $v_e = -2$ V and $v_e = -4$ V. Then, deduce the curve $I = f(v_e)$.

3) The input voltage considered in this case study is sinusoidal ($v_e = 4\sin2\pi Ft$). Draw the representations of the variations of voltage across R_2 in correspondence with the input voltage.

EXERCISE 10

Given is the circuit shown in Figure E10.1.

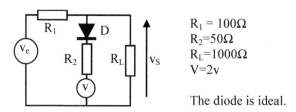

$R_1 = 100\Omega$
$R_2 = 50\Omega$
$R_L = 1000\Omega$
$V = 2v$

The diode is ideal.

Figure E10.1.

1) Input voltage v_e has two values: $v_e = 1$ and 4 V. Determine for each value of v_e the equivalent diagram of the circuit shown in Figure E10.1, as well as the expression of the output voltage v_s.

2) Output voltage ranges between −10 and 10 V and varies in 1 V steps. Represent $v_s = f(v_e)$.

EXERCISE 11

Let us consider the circuit shown in Figure E11.1. Diode D is assumed ideal.

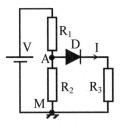

Figure E11.1.

1) Find the equivalent Thévenin generator (R_{TH} and V_{TH}) between points A and M.

2) Determine the expression of current I as a function of V, R_1, R_2 and R_3. Numerical application: V = 10 V; $R_1 = R_2 = 20$ kΩ; $R_3 = 10$ kΩ.

3) Let us now consider the circuit shown in Figure E11.2. Diode D is assumed ideal. Find the Thévenin equivalent generators between B and M and then between N and M. Draw the equivalent diagram under these conditions.

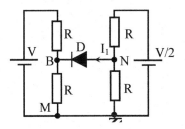

Figure E11.2.

4) Then, deduce the value of current I_1 and the state of diode D.

EXERCISE 12

The circuit to be studied is schematically shown in Figure E12.1. The signal in Figure E12.2 is applied at circuit input.

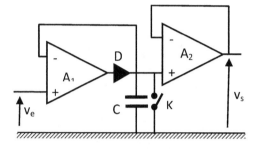

Figure E12.1.

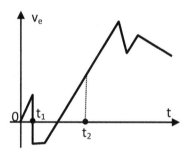

Figure E12.2.

1) On the same diagram, draw the evolution of the output signal v_s in correspondence with signal v_e applied at the input assuming that the switch K is open.

2) The same question as 1), but now the switch K is closed between t_1 and t_2.

EXERCISE 13

The circuit to be studied is presented in Figure E13.1. The voltage applied at input is sinusoidal.

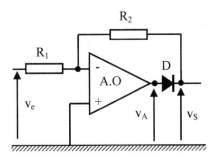

Figure E13.1.

1) Find the expression of the output signal.

2) Represent the evolution in time of voltages v_e and v_s.

3) Find the circuit function when $R_2 = 0$.

4) Let us now consider that $R_2 = R_1 = R$; represent in this case the evolution in time of voltages v_e, v_A and v_s.

5) Represent $v_s = f(v_e)$ and deduce from it the function performed by this circuit.

$$v_e = V_M.\sin(2\pi Ft)$$

EXERCISE 14

Let us consider the circuit shown in Figure E14.1. The input voltage is sinusoidal: $v_e = V_M\sin(\omega t)$.

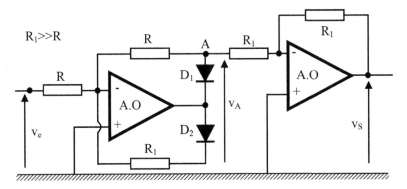

Figure E14.1.

1) Find the definition expressions of voltages v_A and v_s.

2) Represent one below the other the evolutions in time of signals v_e, v_A and v_s.

1.9. Solutions to exercises

SOLUTION TO EXERCISE 1

1) Calculation of the voltage drop across diode (exponential model)

$$I_D = I_1 \cong I_S.(e^{\frac{V_D}{V_T}} - 1) ; \qquad I_D = I_1 \cong I_S.e^{\frac{V_D}{V_T}}$$

$$V_D \cong V_T.Ln(\frac{I_1}{I_S}) ; \qquad V_D \cong V_T.Ln(\frac{I_1}{I_S}) = 539 \text{ mV}$$

2) Expression and value of R in order to impose I_1

$$RI_1 = v_e - V_D; \ R = \frac{v_e - V_D}{I_1} = 1.46 \text{ K}\Omega$$

3) Dynamic resistance R_d of the diode

$$R_d = \frac{dV_D}{dI_D} \qquad I_D = I_1 = I_S.(e^{\frac{V_D}{V_T}} - 1) \equiv I_S.e^{\frac{V_D}{V_T}}$$

Finally, $\dfrac{dI_D}{dV_D} = \dfrac{1}{V_T}I_S.e^{\frac{V_D}{V_T}} = \dfrac{I_1}{V_T}$

Hence, the expression and value of the dynamic resistance R_d of the diode:

$$R_d = \frac{V_T}{I_1} = 26 \ \Omega$$

4) The diode is represented by a simplified model ($V_D = V_0 = 0.6$ V, $R_d = 0$ and R_i is infinite); calculation of current I_2 and comparison with current I_1 (resulting from the use of the exponential model).

Let us recall that $I_1 = 1$ mA

$$I_2 = \frac{v_e - V_D}{R} = \frac{2 - 0.6}{1.46}.10^{-3} = 0.96 \text{ mA}$$

There is a 4% difference with respect to the exponential model. This indicates that the value of I_2 is an approximate value linked to an approximate model.

5) Representation of v_e, v_R and v_S. The diode is represented by the simplified model described at point 4 ($V_D = V_0 = 0.6$ V, $R_d = 0$ and R_i is infinite).

$v_e = 4\sin(2\pi ft+\varphi)$.

When the diode is conducting, the studied circuit is represented by its equivalent diagram in Figure E1.2 (left diagram). On the contrary, when the diode is blocked, the corresponding equivalent diagram is presented in Figure E1.2 (right diagram).

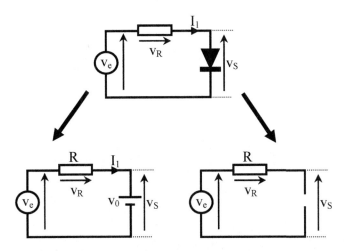

Figure E1.2. *Top: study diagram. Left: equivalent diagram, conducting diode. Right: equivalent diagram, blocked diode*

Consequently, it can be noted that the diode enters in conduction state as soon as the input voltage exceeds the threshold voltage V_0.

It can then be written as: $v_e > V_0$:

$v_s = V_0$ and $v_R = v_s - v_e = V_0 - 4\sin(2\pi ft+\varphi)$

On the contrary, when voltage v_e is lower than the threshold voltage V_0, the diode is blocked. There is no current flow through resistance R.

$v_e < V_0$:

$v_R = 0$

$v_S = v_e - v_R = v_e = 4\sin(2\pi ft+\varphi)$

Figure E1.3 shows the graphic representation of the evolution of various voltages applied to the studied circuit (v_e, v_R and v_S).

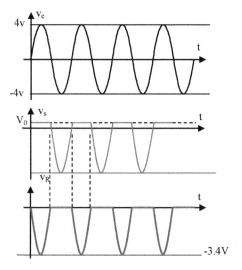

Figure E1.3. *Representation of v_e, v_R and v_S. For a color version of this figure, see www.iste.co.uk/haraoubia/nonlinear1.zip*

SOLUTION TO EXERCISE 2

1) Load line equation.

The load line equation expresses the relation between the current flowing through the diode and the voltage across it:

$$I_D = \frac{v_e - V_D}{R}$$

2) Plot of the direct characteristic of the diode with load line and operating point for the following cases:

2.1) Real diode:

When the diode is represented by the exponential model, Figure E2.4 shows the plot of the characteristic, the static load line and the operating point.

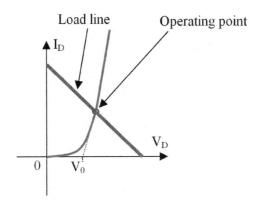

Figure E2.4. *For a color version of this figure, see www.iste.co.uk/haraoubia/nonlinear1.zip*

2.2) Diode in second approximation ($V_0 = 0.6$ V, $R_d = 0$ and R_i is infinite).

For the diode considered in the second approximation, the plot of various curves is shown in Figure E2.5.

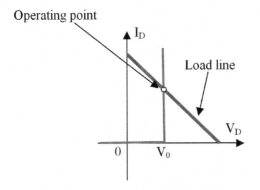

Figure E2.5. *For a color version of this figure, see www.iste.co.uk/haraoubia/nonlinear1.zip*

3) The diode is considered ideal ($V_0 = 0$, $R_d = 0$ and R_i is infinite).

3.1) Expression of $v_s(t)$ and representation of $v_s(t)$:

3.1.1) Expression of $v_s(t)$.

$v_s = 0$ for $v_e > 0$

$v_s = V_m.\sin(2\pi ft)$ for $v_e < 0$

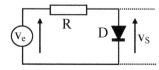

3.1.2) The representation of $v_s(t)$ in correspondence with the input voltage is shown in Figure E2.6.

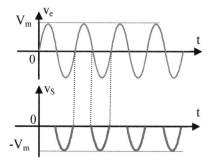

Figure E2.6. *For a color version of this figure, see www.iste.co.uk/haraoubia/nonlinear1.zip*

3.2) Representation (Figure E2.7) of the evolution of output voltage as a function of input voltage $v_s = f(v_e)$.

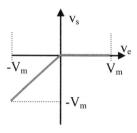

Figure E2.7.

4) The diode is represented by the model shown in Figure E2.8.

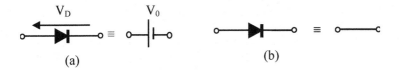

(a)

(b)

Figure E2.8. *Equivalent model of the diode,
a) conducting direction and b) blocked direction*

4.1) Expression of $v_S(t)$, representation of $v_s(t)$:

4.1.1) Expression of $v_S(t)$

$v_S = V_0$ for $v_e > V_0$

$v_S = V_m.\sin(2\pi ft)$ for $v_e < V_0$

4.1.2) Representation of $v_s(t)$

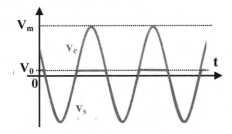

For a color version of this figure, see www.iste.co.uk/haraoubia/nonlinear1.zip

4.2) Representation of $v_s = f(v_e)$

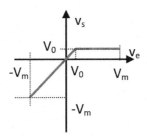

5) The circuit shown in Figure E2.9 is considered. The diode is assumed ideal.

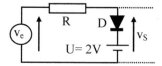

Figure E2.9.

5.1) Filling in the table:

$v_e(v)$	State of the diode	$v_s(v)$
0	Blocked	0
1	Blocked	1
1.5	Blocked	1.5
3	Conducting	2
4	Conducting	2
6	Conducting	2

5.2) Plot of $v_s = f(v_e)$ (see Figure E2.10):

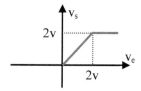

Figure E2.10.

SOLUTION TO EXERCISE 3

1) State of the two diodes, expression of current I across R_L, expression of v_s and calculation of v_s.

1.1) State of diodes D_1 and D_2.

The voltages of diode D_1 at the anode and cathode are 12 and 5 V; therefore, it is conducting.

At the anode of diode D_2, there is a voltage of 8 V, but at its cathode, the voltage is approximately:

$$v_1 - V_0 = (12 - 0.6) = 11.4 \text{ V}$$

Hence, the diode D_2 is blocked and only the diode D_1 is conducting.

1.2) Expression of the current across R_L, expression of v_s and calculation of their value:

1.2.1) Expression of the current across R_L, and numerical value of this current.

When diode D_1 is conducting and diode D_2 is blocked, the equivalent diagram of the studied circuit in Figure E3.1 is shown in Figure E3.2.

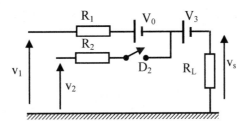

Figure E3.2.

$$I_{R_L} = \frac{v_1 - V_0 - V_3}{R_L + R_1} = \frac{12 - 0.6 - 5}{(10 + 10).10^3} = 0.325.10^{-3} \text{ A} = 325 \, \mu\text{A}$$

1.2.2) Expression of v_s and numerical value of v_s.

$$v_s = R_L.I_{R_L} = R_L \frac{v_1 - V_0 - V_3}{R_L + R_1} = 10^4 \frac{12 - 0.6 - 5}{(10 + 10).10^3} = 3.25 \text{ V}$$

2) Voltages v_1 and v_2 are sinusoidal, expression of the output voltage v_s and time representation of v_s.

2.1) Expression of v_s:

$$v_1 = 12\sin(2\pi ft) \text{ and } v_2 = 8\sin(2\pi ft) \text{ and } V_3 = 5 \text{ V}.$$

It should be noted that only diode D_1 is conducting and D_2 is blocked, since v_1 is always above v_2. As long as v_1 is below $V_3 + V_0$, diodes D_1 and D_2 are blocked and voltage v_s is zero: $v_s = 0$

As soon as v_1 exceeds voltage $(V_3 + V_0)$, diode D_1 becomes conducting and diode D_2 remains blocked. Then:

$$v_s = R_L \frac{v_1 - V_0 - V_3}{R_L + R_1}; \quad v_s = 6\sin(2\pi ft) - 2.8$$

2.2) Time representation of v_s (see Figure E3.3):

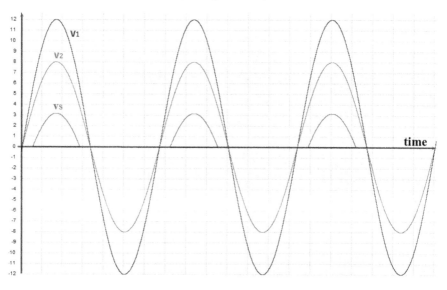

Figure E3.3. *For a color version of this figure, see*
www.iste.co.uk/haraoubia/nonlinear1.zip

SOLUTION TO EXERCISE 4

1) Expression of the value of voltage v_3 and value of the current across resistance R_3:

$v_1 = 1$ V, $v_2 = 4$ V and $R_2 = R_3 = 1$ kΩ.

Under these conditions, diode D is blocked and the equivalent circuit is shown in Figure E4.3.

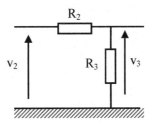

Figure E4.3.

The expression of voltage v_3 is:

$$v_3 = v_2 \frac{R_3}{R_2 + R_3} \; ; \qquad v_3 = \frac{v_2}{2} = 2 \text{ V}$$

$$I_{R3} = \frac{v_3}{R_3} = 2 \text{ mA}$$

2) Expression and value of voltage v_3 and value of the current across resistance R_3 when $v_1 = 3$ V, $v_2 = 3$ V and $R_1 = R_2 = R_3 = 1$ kΩ.

Under these conditions, the diode enters a conduction state and is equivalent to its threshold voltage V_0. The equivalent diagram in Figure E4.4 is obtained.

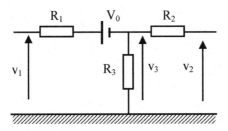

Figure E4.4.

Applying Millman's theorem or Kirchoff's theorem, the following relation is obtained:

$$v_3 = \frac{(v_1 - V_0)R_2 R_3 + v_2 R_1 R_3}{R_1 R_2 + R_1 R_3 + R_2 R_3}$$

When $v_1 = 4$ V, $v_2 = 4$ V and $R_1 = R_2 = R_3 = 1$ kΩ

$$v_3 = \frac{(v_1 - V_0) + v_2}{3} = \frac{4 - 0.6 + 4}{3} = 2.46 \text{ V}$$

The current across R_3 is defined as:

$$I_{R3} = \frac{v_3}{R_3} = \frac{2.46}{10^3} = 2.46 \text{ mA}$$

3) Representation of the signal $v_3(t)$. Points A and B of the circuit shown in Figure E4.1 of exercise 4 are joined.

Furthermore, $R_2 = 5$ kΩ and $R_1 = R_3 = 1$ kΩ. The signal s(t) applied at points A and B is triangular.

Let us recall that when diode D enters the conduction state $s(t) > V_0$, the expression of signal $v_3(t)$ is given by:

$$v_3(t) = \frac{(v_1(t) - V_0)R_2R_3 + v_2(t)R_1R_3}{R_1R_2 + R_1R_3 + R_2R_3}$$

Since $v_1(t)$ and $v_2(t)$ are identical ($v_1 = v_2 = s(t)$), $R_2 = 5$ kΩ; $R_1 = R_3 = 1$ kΩ and $V_0 = 0.6$ V, the following can be written:

$$v_3(t) = \frac{(v_1(t) - V_0)R_2R_3 + v_2(t)R_1R_3}{R_1R_2 + R_1R_3 + R_2R_3} = \frac{6v_1(t) - 3}{11} = \frac{6s(t) - 3}{11}$$

When the diode is blocked, $v_3(t)$ is defined as:

$$v_3(t) = s(t)\frac{R_3}{R_2 + R_3} = \frac{s(t)}{6}$$

Hence, the evolution of voltage $v_3(t)$ in correspondence with signal s(t) is shown in Figure E4.5.

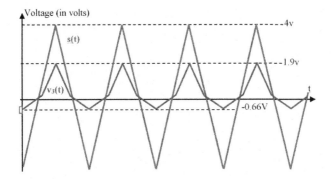

Figure E4.5. *For a color version of this figure, see www.iste.co.uk/haraoubia/nonlinear1.zip*

SOLUTION TO EXERCISE 5

1) Representation of the characteristic (Figure E5.3) of an ideal diode:

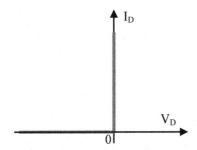

Figure E5.3. *Characteristic $I_D = f(V_D)$ of an ideal diode. For a color version of this figure, see www.iste.co.uk/haraoubia/nonlinear1.zip*

2) Expression and value of the maximum current across diode D:

$$I_{max} = \frac{V}{R_1} \; ; I_{max} = 12 \text{ mA}$$

3) Equation of the load line and its representation in the same diagram with the characteristic:

$$I_D = (V - V_D)/R$$

The specific points of the load line are the following:

point A: $V_D = 0$; $I_D = (V/R) = 12$ mA; point B: $I_D = 0$; $V_D = V$.

Representation (see Figure E5.4):

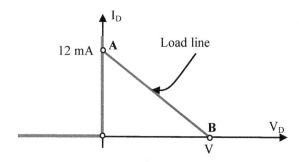

Figure E5.4. *Direct characteristic $I_D = f(V_D)$ and load line. For a color version of this figure, see www.iste.co.uk/haraoubia/nonlinear1.zip*

4) Filling in of Table E5.2 (Figure E5.2; D_1 and D_2 are assumed ideal).

v_1	v_2	State D_1	State D_2	v_{S1}	v_{S2}
2v	5v	B	P	2v	5v
0v	2v	B	P	0v	2v
−4v	−2v	B	P	−4v	−2v
4v	2v	P	B	2v	4v
−1v	−3v	P	B	−3v	−1v
4v	0v	P	B	0v	4v

Table E5.2.

5) Representation of v_{S1} and v_{S2} when v_1 is sinusoidal with peak amplitude equal to 3 V and $v_2 = 2$ V.

$$v_1 = 3\sin(2\pi ft)$$

It is sufficient to note that when $v_1 > v_2$, diode D_1 is conducting and diode D_2 is blocked and therefore:

$$v_{S1} = v_2 \text{ and } v_{S2} = v_1.$$

On the contrary, when $v_1 < v_2$, diode D_1 is blocked and diode D_2 is conducting; then, $v_{S1} = v_1$ and $v_{S2} = v_2$.

Hence, v_{S1} and v_{S2} are represented in Figure E5.5.

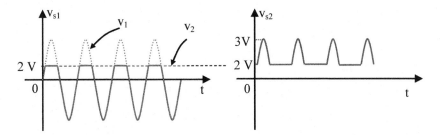

Figure E5.5. *Representation of $v_{S1}(t)$ and $v_{S2}(t)$. For a color version of this figure, see www.iste.co.uk/haraoubia/nonlinear1.zip*

SOLUTION TO EXERCISE 6

1) Expression of voltage across R_1.

When $v_1 < 2$ V, diode D_1 is blocked and D_2 is conducting:

$$V_{R1} = E = 2 \text{ V}$$

When $v_1 > 2$ V, diode D_1 is conducting and D_2 is blocked:

$$V_{R1} = v_1$$

2) Representation (Figure E6.2) of the evolution of $v_{R1}(t)$:

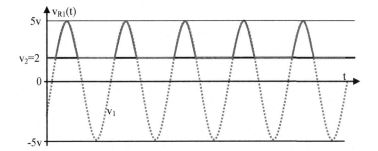

Figure E6.2. *Evolution of $v_{R1}(t)$ in time. For a color version of this figure, see www.iste.co.uk/haraoubia/nonlinear1.zip*

SOLUTION TO EXERCISE 7

1) Calculation of V_{T1}:

$V_T = KT/e$; $V_{T1} = KT_1/e$ and $V_{T0} = KT_0/e$;

$V_{T1} = (V_{T0}.T_1/T_0)$; A.N: $V_{T1} = 30.3$ mV

2) Calculation of reverse saturation current I_{S1}:

$$I_D = I_{S1}\left[e^{\frac{V_D}{V_{T1}}} - 1\right]; \quad I_{S1} = \frac{I_D}{\left[e^{\frac{V_D}{V_{T1}}} - 1\right]}; \quad \text{A.N: } I_{S1} \cong 10^{-6} \text{ A} = 1 \text{ μA}$$

Reverse current I_{S1} remains very small compared to I_D even though temperature is above 50°C with respect to ambient temperature. It can be negligible with respect to I_D under all operating conditions.

3) Calculation of reverse saturation current I_{S0} at $T_0 = 300$ K and comparison between I_{S1} and I_{S0}.

$$I_{S0} = \frac{I_S}{\left[e^{a(T-T_0)}\right]};$$

A.N: I_{S0} at $T_0 = 300$ K; $I_{S0} = 49.78.10^{-9}$ A $= 49.78$ nA,

$I_{S1} \gg I_{S0}$; I_S increases when temperature T rises

4) Calculation of temperature variation ΔT required for the reverse saturation current I_S to double its value.

$$I_S = I_{S0}\left[e^{a(T-T_0)}\right]; \quad I_S = 2 \, I_{S0};$$

$$\left[e^{a(T-T_0)}\right] = 2; \quad \left[e^{a\Delta T}\right] = 2;$$

A.N: $\Delta T = \ln(2)/a = 11.55$ °C

SOLUTION TO EXERCISE 8

1) Representation of voltages v_{R1}, v_{R2} and v_D and currents I_1, I_2 and I_D (Figure E8.4):

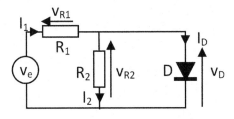

Figure E8.4.

2) Voltage v_e from which diode D conducts.

The diode conducts from voltage equal to its threshold voltage:

$v_D \geq 0.6$ V;

$v_D = v_{R2} \geq 0.6$ V.

Just before the diode conducts, the following can be written:

$$v_{R2} = \frac{R_2}{R_1 + R_2} v_e = \frac{v_e}{2}$$

The diode conducts provided that:

$$v_{R2} \geq 0.6V \quad \Rightarrow \quad \frac{1}{2}v_e \geq 0.6V \ ;$$

$v_e \geq 1.2$ V

3) Equation of the load line, the plot and the coordinates of the operating point.

3.1) Equation of the load line:

$$I_D = \frac{v_{e1} - v_D}{R_1}$$

3.2) Plot of the load line (Figure E8.5):

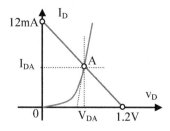

Figure E8.5. *For a color version of this figure, see*
www.iste.co.uk/haraoubia/nonlinear1.zip

3.3) Coordinates of the operating point A (I_{DA}, V_{DA})

$$v_{DA} \cong V_0$$

$$I_{DA} = \frac{v_{e1} - v_{DA}}{R_1} = \frac{v_{e1} - V_0}{R_1}$$

Threshold voltage $V_0 \cong 0.6$ V

$$I_D = 6.10^{-3} = 6 \text{ mA}$$

4) Representation of v_{e1}, v_{R1} and v_D (Figure E8.6):

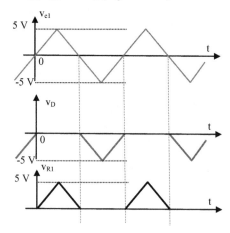

Figure E8.6. *For a color version of this figure, see*
www.iste.co.uk/haraoubia/nonlinear1.zip

SOLUTION TO EXERCISE 9

1) Calculation of the current I across the diode for the following cases: $v_e = 2$ V, $v_e = 4v$, $v_e = -0.5v$; $v_e = -2v$ and $v_e = -4v$ and curve: $I = f(v_e)$.

1.1) Current across the diode:

In order to simplify the circuit, Thévenin's theorem is applied between points A and M (Figure E9.2):

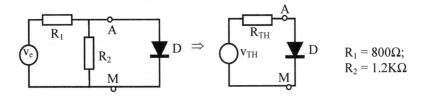

$R_1 = 800\Omega$;
$R_2 = 1.2K\Omega$

Figure E9.2.

$$v_{TH} = v_e \frac{R_2}{R_1 + R_2} = 0.6 V_e$$

$$R_{TH} = \frac{R_1 . R_2}{R_1 + R_2} = 480 \ \Omega$$

When the diode is blocked, I=0.

When the diode conducts:

$$I = \frac{v_{TH} - v_D}{R_{TH}} = \frac{v_{TH}}{R_{TH}} = 0.6 \frac{v_e}{R_{TH}}$$

The various values of current I as well as the operating state of diode D depending on the value of input voltage are given in Table E9.1.

v_e (v)	−4	−2	0.5	2	4
v_{TH} (v)	−2.4	−1.2	0.3	1.2	2.4
State of the diode	Blocked	Blocked	Conducting	Conducting	Conducting
$I = \dfrac{V_{TH}}{R_{TH}}$	0	0	625 μA	2.5 mA	5 mA

Table E9.1.

1.2) Representation of the variations of $I = f(v_e)$ (see Figure E9.3):

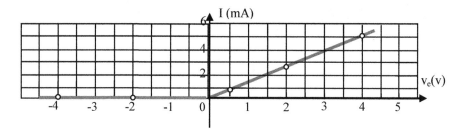

Figure E9.3. *Representation of $I = f(v_e)$*

2) Calculation of the current I across the diode ($V_0 = 0.6$ V) when: $v_e = 4$ V, $v_e = 2$ V, $v_e = 1$ V, $v_e = −0.5$ V, $v_e = −2$ V and $v_e = −4$ V. Graphical representation of $I = f(v_e)$:

2.1) Calculation of current I:

$$I = \frac{V_{TH} - V_D}{R_{TH}}$$

$$I = \frac{0.6(V_e - 1)}{R_{TH}}$$

The values of current I depending on the value of v_e are summarized in Table E9.2.

v_e (v)	−4	−2	0.5	1	2	4
v_{TH} (v)	−2.4	−1.2	0.3	0.6	1.2	2.4
State diode	B	B	B	B	C	C
$I = \dfrac{V_{TH} - 0.6}{R_{TH}}$	0	0	0	0	1.25 mA	3.75 mA

Table E9.2. *(B: blocked; C: conducting)*

2.2) Representation of current I across the diode depending on voltage input (see Figure E9.4):

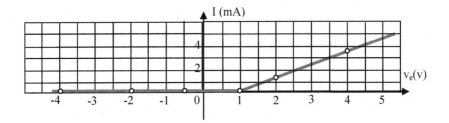

Figure E9.4.

3) v_e is a sinusoidal voltage ($v_e = 4\sin2\pi Ft$). The diode has a threshold equal to 0.6 V. Voltage variations across R_2 (see Figure E9.5):

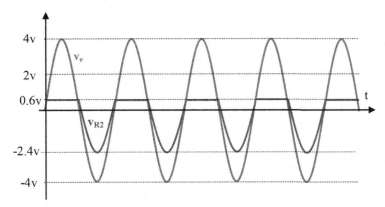

Figure E9.5. *Evolution of $v_{R2}(t)$ in correspondence with $v_e(t)$. For a color version of this figure, see www.iste.co.uk/haraoubia/nonlinear1.zip*

SOLUTION TO EXERCISE 10

1) Equivalent diagram of the circuit, expression and value of v_s.

1.1) Case of $v_e = 1$ V:

The voltage ($v_e = 1$ V) applied at the anode of diode D is below the voltage applied at its cathode ($v = 2$ V).

The diode is blocked, and the equivalent diagram is shown in Figure E10.2:

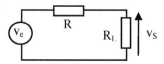

Figure E10.2. *Equivalent diagram of the circuit in Figure E10.1 when the diode is blocked*

$$v_S = \frac{R_L}{R_L + R_1}.v_e = 0.91 \text{ V}$$

1.2) Case of $v_e = 4$ V.

Diode D conducts, and the equivalent diagram of the circuit studied is shown in Figure E10.3.

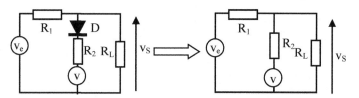

Figure E10.3. *Equivalent diagram of the circuit in Figure E10.1 when the diode is conducting*

$$v_S = \frac{R_L R_1 v + R_L R_2 v_e}{R_1 R_2 + R_1 R_L + R_2 R_L}, \quad v_S = 2.58 \text{ Volts}$$

2) Representation of $v_s = f(v_e)$ (see Figure E10.4):

$v_e \leq 2$ The diode is blocked: $v_S = \frac{R_L}{R_L + R_1}.v_e = 0.91.v_e$

$v_e \geq 2$ conducting diode:

$$v_S = \frac{R_L(R_1 v + R_2 v_e)}{R_1 R_2 + R_1 R_L + R_2 R_L} \; ; \; v_S = 0.3225.v_e + 1.290$$

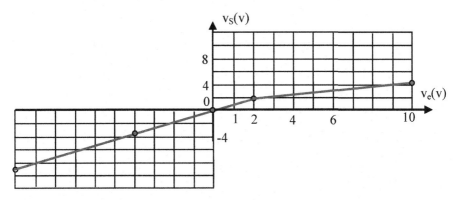

Figure E10.4. *Representation of $v_s = f(v_e)$. For a color version of this figure, see www.iste.co.uk/haraoubia/nonlinear1.zip*

SOLUTION TO EXERCISE 11

1) Equivalent Thévenin generator (Figure E11.3):

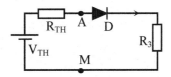

Figure E11.3. *Diagram of equivalent Thévenin generator of the circuit shown in Figure 11.1*

$$R_{TH} = \frac{R_1 R_2}{R_1 + R_2} = 10 \text{ K}\Omega$$

$$V_{TH} = \frac{R_2}{R_1 + R_2} V = \frac{V}{2} = 5 \text{ V}$$

2) Expression of current I as a function of V, R_1, R_2 and R_3, and numerical application.

The diode D is considered ideal: $V_0 = 0$ and $R_d = 0$, and reverse resistance is assumed infinite.

$$I = \frac{V_{TH}}{R_{TH} + R_3} = \frac{V \dfrac{R_2}{R_1 + R_2}}{\dfrac{R_1 R_2}{R_1 + R_2} + R_3} = \frac{V.R_2}{R_1 R_2 + R_1 R_3 + R_2 R_3}$$

Numerical application: $V = 10$ V; $R_1 = R_2 = 20$ kΩ; $R_3 = 10$ kΩ; $I = 250$ μA

3) Equivalent Thévenin generators between B and M and then between N and M and equivalent diagram:

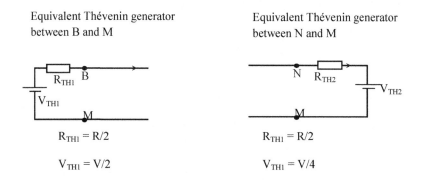

Equivalent Thévenin generator between B and M

Equivalent Thévenin generator between N and M

$R_{TH1} = R/2$ $R_{TH1} = R/2$

$V_{TH1} = V/2$ $V_{TH1} = V/4$

4) Deduction of the value of current I_1.

The overall equivalent diagram of the circuit shown in Figure E11.2 is presented in Figure E11.4.

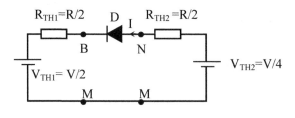

Figure E11.4.

Since $V_{TH2} < V_{TH1}$, diode D is blocked. There is no current across it:

I = 0

SOLUTION TO EXERCISE 12

1) Evolution of the output signal v_s in correspondence with the signal v_e when the switch K is open (see Figure E12.3).

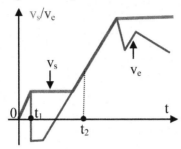

Figure E12.3. *Evolution of the output voltage in correspondance with input voltage. For a color version of this figure, see www.iste.co.uk/haraoubia/nonlinear1.zip*

The capacitor charges rapidly through the dynamic resistance of the diode with a time constant that is practically zero when the diode is in conduction state.

When the diode is blocked, the capacitor cannot discharge, given the high input resistance of the follower circuit (follower circuit A_2).

2) Evolution of the output signal v_s in correspondence with signal v_e when the switch K is closed between instants t_1 and t_2 (see Figure E12.4).

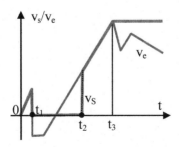

Figure E12.4. *Evolution of the output voltage in correspondance with input voltage. For a color version of this figure, see www.iste.co.uk/haraoubia/nonlinear1.zip*

The voltage across capacitor (v_c) and consequently the output voltage (v_s) follow the ascending form of v_e. Between t_1 and t_2, the voltage across the capacitor is zero (short-circuit capacitor): $v_s = 0$. Starting from t_2, the voltage v_s instantaneously takes the value of v_e until t_3. From this instant on, the voltage maintains the value $v_e(t_3)$, since the diode is blocked and the capacitor cannot discharge, given the high impedance of the follower operational amplifier.

SOLUTION TO EXERCISE 13

1) Expression of the output signal:

Since the input voltage is sinusoidal, this problem is approached by separating the positive half-wave from the negative half-wave.

1.1) Case of $v_e > 0$:

The diode is blocked: $\mathbf{v_s = v_e}$.

1.2) Case of $v_e < 0$:

The diode is conducting: $v_s = \dfrac{-R_2}{R_1} v_e$

2) Representation (Figure E13.2) of the evolution in time of voltages v_e and v_s.

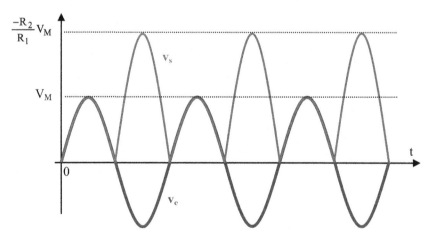

Figure E13.2. *Evolution of v_s in correspondence with v_e. For a color version of this figure, see www.iste.co.uk/haraoubia/nonlinear1.zip*

3) Function of the circuit when $R_2 = 0$.

When $R_2 = 0$, the following expressions of the output voltage are found:

$v_e > 0$: $v_s = v_e$

$v_e < 0$: $v_s = 0$

The circuit assures the function of half-wave rectifier.

4) Evolution of voltages v_e, v_A and v_s when $R_2 = R_1 = R$ (see Figure E13.3):

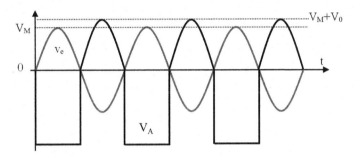

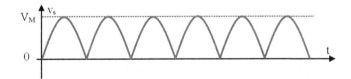

Figure E13.3. *Representation of v_e, v_A and v_s. For a color version of this figure, see www.iste.co.uk/haraoubia/nonlinear1.zip*

4.1) Case of $v_e > 0$.

The diode is blocked: $v_s = v_e$ and $v_A = -V_{cc}$, $\pm V_{cc}$ is the voltage of symmetric supply of the operational amplifier

4.2) Case of $v_e < 0$.

The diode is conducting: $v_s = -v_e$ and $v_A = -v_e + V_0$, where V_0 is the threshold voltage of the diode.

5) Representation of $v_s = f(v_e)$ and function fulfilled by the circuit.

Let us recall that when voltage v_e is positive, $\mathbf{v_s} = \mathbf{v_e}$, and when voltage v_e is negative, $\mathbf{v_s} = -\mathbf{v_e}$

Thus, the representation of v_s as a function of v_e is given in Figure E13.4:

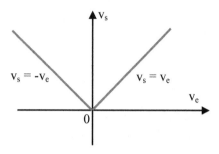

Figure E13.4. *Representation of v_s as a function of v_e*

The function performed by the circuit is full-wave rectification without thresholds.

SOLUTION TO EXERCISE 14

1) Definition expressions of voltages v_A and v_s:

1.1) $v_e > 0$

Under these conditions, diode D_1 is conducting and diode D_2 is blocked. Therefore, the following can readily be noted:

$$v_A = -v_e \text{ and } v_s = -v_A = v_e$$

1.2) $v_e < 0$

Diode D_1 is blocked and diode D_2 is conducting:

$$v_A = v_e \text{ and } v_s = -v_A = -v_e$$

2) Representation as a function of time of signals v_e, v_A and v_s (see Figure E14.2)

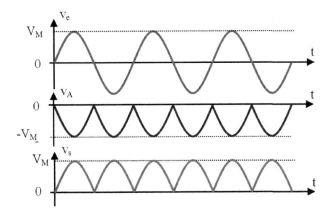

Figure E14.2. *Representation as a function of time of signals v_e; v_A; v_s.*
For a color version of this figure, see www.iste.co.uk/haraoubia/nonlinear1.zip

Low-frequency Oscillators

2.1. Feedback study

Whenever the output influences the input, there is feedback. This involves loop systems, as indicated by the diagram shown in Figure 2.1.

Returning of a fraction of the signal from the output to the input constitutes a "feedback" assembly.

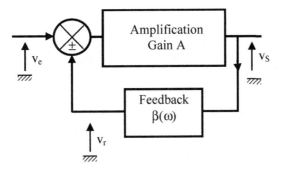

Figure 2.1. *Feedback loop system*

Feedback can be positive or negative. Positive feedback involves adding to the input a fraction of the output voltage.

Negative feedback allows taking out a fraction of the output voltage returned by the feedback circuit from the input voltage "v_e".

2.1.1. *Negative feedback*

The assembly shown in Figure 2.2 is used for the study of negative feedback. It can be seen that the part of the output voltage "v_s" that is returned to the input is taken out from the input voltage "v_e". Under these conditions, the following can be written:

$$v_1 = v_e - v_r = \frac{v_S}{A}$$

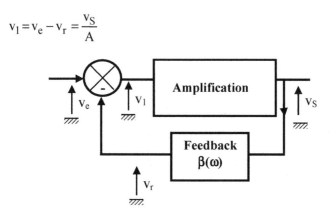

Figure 2.2. *Organization of negative feedback*

Voltage "v_r" returned from the output to the input is out of phase with the input voltage "v_e". Therefore, these two signals are subtracting their effects. They produce an output signal that is less than that existing in an open loop.

$$v_r = \beta.v_S; \quad v_e - \beta v_S = \frac{v_S}{A} \quad \Rightarrow \quad \frac{v_S}{v_e} = \frac{A}{1 + A\beta}$$

Negative feedback is also called inverse feedback. Its role is to modify the performances of a system (e.g. an amplifier).

The properties of a negative feedback amplifier are modified as follows:

– stabilization of the gain value;

– reduction of nonlinear distortion;

– reduction of the internal noise effect;

– increase of the amplifier bandwidth with respect to its open-loop bandwidth;

– control of the value of input and output impedances.

Thus, the superposition of the input signal and of a fraction of the output signal leads to modified performances of the amplifier or of any other device.

2.1.2. *Positive feedback*

In a device with a positive feedback loop, the part of voltage returned from the output to the input is added to the input voltage, as shown in Figure 2.3.

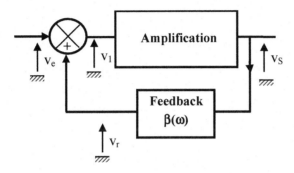

Figure 2.3. *Organization of positive feedback*

This system may very rapidly become unstable, since with each feedback loop, a part of the output voltage is added to the input voltage and amplified.

$$v_1 = v_r + v_e$$

The following is the relation between input and output:

$$\frac{v_S}{v_e} = \frac{A}{1 + A\beta}$$

This process leads to a destabilization of the output signal. This effect is sought for by a great number of electronic devices such as oscillators, comparators, astable multivibrators, or triggers.

2.1.3. *Oscillators and positive feedback*

A very relevant example, namely that of oscillators, will be studied subsequent to this presentation.

An oscillator is by definition a device that generates on its own a periodic alternative output voltage without any input excitation.

The device using a positive feedback can be modified and, under a certain condition, can operate as an electrical oscillator, as shown in Figure 2.4.

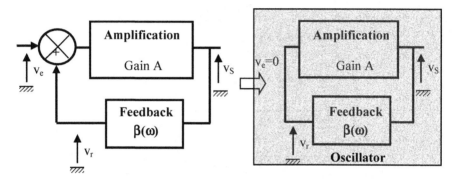

Figure 2.4. *Positive feedback in an electrical oscillator*

The generated wave can be sinusoidal, rectangular or triangular. Oscillators can deliver voltages with very broad frequency ranges, from a few tens of hertz to tens of gigahertz.

Thus, depending on the shape of the wave to be generated or the expected frequency, there can be radically different approaches to the study and design of an oscillator.

Oscillators can be classified into two large families depending on the shape of the wave they generate, as shown in Figures 2.5(a) and 2.5(b).

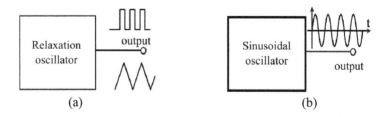

Figure 2.5. *Principle of a) a relaxation oscillator; b) a sinusoidal oscillator*

Devices that deliver waves with a frequency range rich in harmonics are called "relaxation oscillators".

On the contrary, systems that generate waves containing only one harmonic are called "sinusoidal oscillators".

2.2. Principle of sinusoidal feedback oscillator

Irrespective of its type, the block diagram of a sinusoidal feedback oscillator is shown in Figure 2.6. The feedback network is a passive dissipative oscillating circuit. Its role is to fix the oscillating frequency. Left on its own, this network (feedback network) cannot produce sustained oscillations.

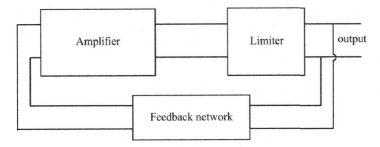

Figure 2.6. *Block diagram of a sinusoidal feedback oscillator*

Thus, in order to produce and sustain an oscillation, this network must be supplied with external energy. This role can be ensured by an amplifier circuit, for example. The limiter circuit stabilizes the amplitude of the output signal at a well-established limit.

2.3. Oscillator parameters

The study of sinusoidal oscillators is a delicate task. In the most general case, an oscillator is described by a differential equation, the order of which is greater than or equal to two. In order to design and realize an oscillator, the following four parameters should be thoroughly defined:

1) oscillation frequency;

2) condition for sustained oscillation;

3) amplitude of the output signal;

4) distortion factor of the output signal.

The first two parameters (oscillating frequency and condition for sustained oscillation) can be defined using the linear theory. This is sufficient for a brief description of an oscillator. When accurate information on the amplitude and the distortion factor of the output signal is required, nonlinear theory should be employed. Herein, the focus is on how to determine the oscillation frequency and the condition to sustain it.

The nonlinear aspect of oscillators will be addressed at the end of the Chapter 3, which is dedicated to high-frequency oscillators.

2.4. Linear mode oscillator operation

2.4.1. *Introduction*

To deliver sustained oscillations, an oscillator must receive external energy. The latter can be supplied by one of the following elements:

– amplification of the damped oscillating signal;

– external voltage source;

– negative resistance.

An amplifier element is very often used in order to sustain low- or high-frequency oscillations. The type of oscillator circuit that uses an amplifier circuit and an oscillating circuit on a feedback loop is called a "feedback oscillator". In this chapter, we study this device in the field of low frequencies. There is no interest in using an external source for oscillators. Oscillators with negative resistance generally employ the specific characteristics of certain diodes (e.g. tunnel diodes) to sustain sinusoidal oscillations. Furthermore, negative resistances can be artificially realized by means of active components, such as operational amplifiers.

2.4.2. *Feedback oscillator*

A feedback oscillator is essentially constituted of two quadripoles assembled as shown in Figure 2.7.

The amplification element takes into account the nonlinearity aspect previously represented by an amplitude limiter.

As already mentioned, the study is conducted under linear conditions (steady state). This means that the nonlinear element has already performed its task and it consequently introduces only a constant coefficient included in the amplification term. Its existence can be completely ignored in this steady-state (linear conditions) context.

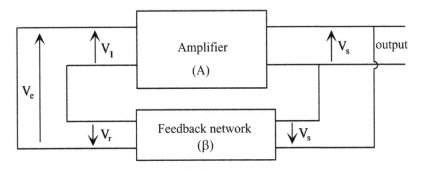

Figure 2.7. *Principle of feedback oscillator*

In Figure 2.7, v_r is the voltage returned from output to input in order to sustain the oscillations; v_s is the output voltage applied to the load; v_1 is the input voltage applied to the amplifier and v_e is the voltage applied at oscillator input ($v_e = 0$ for an oscillator).

An oscillator is an unstable system. Let us now write the analytical equations defining the operation of the oscillator and find under what condition the latter becomes unstable.

$$v_r = \beta . v_s; \; v_1 = (v_s/A); \; v_e = -v_r + v_1; \; v_e = -\beta . V_s + (V_s/A)$$

$$\frac{v_s}{v_e} = \frac{A}{1 - A\beta}$$

To reach system stability, according to Nyquist, the condition $A.\beta < 1$ should be met. It is our objective to render the system fully unstable. For this purpose, let us have $A.\beta = 1$.

Under these conditions, $\dfrac{v_s}{v_e} \to \infty$

This is equivalent to voltage v_e (supposed input voltage of the system) being zero (see Figure 2.7) and the existence of an output voltage v_s. This is how a voltage v_s is generated without applying any input voltage. For the oscillation to be sinusoidal, it is sufficient to use the feedback circuit (of transfer function β) to select a properly given harmonic.

IN SUMMARY.–

A device such as the one shown in Figure 2.7 operates as an oscillator provided that A.$\beta = 1$.

This is called the Barkhausen condition and, in reality, it covers two conditions: one relating to the phase and the other to the module.

$$A.\beta \Rightarrow \begin{array}{c} \text{Arg}(A) + \text{Arg}(\beta) = 0 \quad [2\pi] \\ |A.\beta| = 1 \end{array}$$

The condition imposed on the argument allows the determination of the oscillation frequency (f_0) of the circuit. The condition imposed on the module leads to determining the condition for sustained oscillation.

Gain A is generally real, which is the case of low-frequency amplifiers.

The Barkhausen condition comes down to:

$$A.\beta \Leftrightarrow \begin{array}{c} \text{Arg}(\beta) = 0 \quad [2\pi]; \quad \text{if} \quad A > 0 \\ |A.\beta| = 1 \end{array}$$

$$A.\beta \Leftrightarrow \begin{array}{c} \text{Arg}(\beta) = \pi \quad [2\pi]; \quad \text{if} \quad A < 0 \\ |A.\beta| = 1 \end{array}$$

If the argument of (β) is zero or equal to (π), then the imaginary part of the transfer function β must necessarily be zero.

When the imaginary part of β is zero, the expression of the oscillation frequency f_0 can be deduced.

The condition on the module can be reduced to:

$$A(f_0).\beta(f_0) = 1 \Leftrightarrow A.\beta(f_0) = 1 \text{ (if A is real)}$$

2.5. Phase-shift oscillators

2.5.1. *Schematic diagram and equation*

This type of oscillator is constituted of an amplifier and a certain number of "RC" cells. Most commonly, three cells are used (Figure 2.8) for the following two essential reasons:

1) three cells easily allow a phase rotation of 180°. This result cannot be obtained with only two cells; and

2) a phase rotation of 180° can be obtained with four cells or even more, but this increases the number of components, and the overall size is bigger than that for three cells.

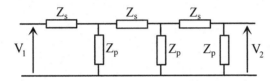

Figure 2.8. *Cells in the phase shift network*

The phase rotation of 180° is required due to use in the chain that comprises the oscillator of a real gain amplifier.

Thus, using three low-pass or high-pass cells connected to an amplifier, a low-frequency phase-shift oscillator with reduced overall size can be realized, as suggested in Figure 2.9.

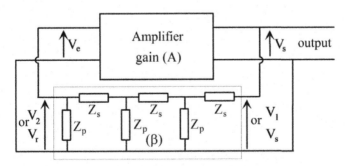

Figure 2.9. *Schematic diagram of the phase shift oscillator*

To operate as an oscillator, this circuit must fulfill the Barkhausen condition:

$$A.\beta = 1 \quad \Rightarrow \quad Arg(A) + Arg(\beta) = 0 + 2k\pi \text{ and } |A.\beta| = 1$$

where A is the gain introduced by the amplifier and β is the transfer function of the feedback circuit represented by the phase-shift network.

As a first step, and for the sake of simplicity, let us suppose that the input impedance of the amplifier is very high. Consequently, it does not charge the feedback circuit. To ease calculation, the circuit can be opened at point N, as shown in Figure 2.10.

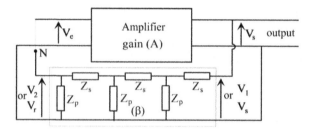

Figure 2.10. *Opening of the circuit at point N in order to separate the amplifier and feedback circuits*

This enables the separate calculation of the amplifier gain A and of the expression of transfer function β.

$$A = \frac{V_s}{V_e}; \quad \beta = \frac{V_r}{V_1} = \frac{V_r}{V_s}$$

NOTE.– If the input impedance of the amplifier is not high, it is imperative to restore the impedance balance after cutting off, but before making all the calculation. The separation of amplifier and feedback circuits facilitates calculation, especially for the transfer function of the feedback circuit.

For the calculation of β, let us use the most classic method, which involves writing the equations for the three circuit meshes in Figure 2.11.

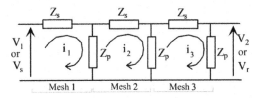

Figure 2.11. *Feedback phase-shift circuit for the calculation of β*

Mesh 1: $v_s = Z_s.i_1 + Z_p.(i_1 - i_2)$　　　　　　　　　　　　　　　[2.1]

Mesh 2: $0 = Z_p.(i_2 - i_1) + Z_s.i_2 + Z_p.(i_2 - i_3)$　　　　　　　　[2.2]

Mesh 3: $0 = Z_p.(i_3 - i_2) + Z_s.i_3 + Z_p.i_3$　　　　　　　　　　　[2.3]

And also: $i_3 = \dfrac{v_r}{Z_p}$　　　　　　　　　　　　　　　　　　　[2.4]

The system of equations allows the determination of the (v_r/v_s) ratio. In equation [2.3], i_3 is replaced by its equivalent expression and then i_2 is deduced:

$$(2.Z_p + Z_s)i_3 = Z_p\, i_2 \quad \Rightarrow \quad i_2 = \frac{(2Z_p + Z_s)v_r}{Z_p^2}$$

From equation [2.2], the expression of i_1 is deduced as a function of v_r:

$$\left[(2Z_p + Z_s)^2 \frac{v_r}{Z_p^2}\right] - v_r = Z_p i_1$$

$$i_1 = \left[\frac{(2Z_p + Z_s)^2}{(Z_p)^2} - 1\right]\frac{v_r}{Z_p}$$

This leads to deducing the expression of the output voltage as a function of the voltage returned to the input.

$$v_s = \left[Z_p + Z_s\right].\left[\frac{(2Z_p + Z_s)^2}{(Z_p)^2} - 1\right]\frac{v_r}{Z_p} - \frac{(2Z_p + Z_s)}{Z_p}v_r$$

Finally:

$$\beta = \frac{v_r}{v_s} = \frac{1}{\left[\dfrac{Z_s}{Z_p}\right]^3 + 5\left[\dfrac{Z_s}{Z_p}\right]^2 + 6\left[\dfrac{Z_s}{Z_p}\right] + 1}$$

As already noted, RC cells can be of a high-pass or low-pass type. Two cases can therefore be presented for impedances Z_p and Z_s:

1) purely reactive Z_p and purely resistive Z_s (low-pass cells); and

2) purely resistive Z_p and purely reactive Z_s (high-pass cells).

2.5.2. Low-pass cells

The series impedance comprises a resistance, and the parallel impedance comprises a capacitance:

$$Z_s = R \text{ and } Z_p = (1/jC\omega)$$

2.5.2.1. Calculation of the oscillation frequency

For the sake of simplicity, let us suppose that the low-frequency amplifier element has a purely real gain contribution to the working frequency (which is generally the case in practice for low frequency).

The condition on the argument related to the oscillator operation is defined as follows:

$$\text{Arg}(A) + \text{Arg}(\beta) = 0 \ [2\pi]$$

Since the amplifier gain A has been supposed real, the transfer function β must necessarily also be real for a certain frequency f_0 (oscillation frequency). The transfer function β has already been expressed. In the present case, replacing Z_s and Z_p by their equivalent relation is sufficient:

$$\beta = \frac{v_r}{v_s} = \frac{1}{-j[RC\omega]^3 - 5[RC\omega]^2 + 6j[RC\omega] + 1}$$

β must be real. Its imaginary part must therefore be zero. The annulation of this imaginary part leads to finding the expression of the oscillation frequency f_0 of the concerned oscillator:

$$Im^{*}(\beta) = RC\omega_0[6 - (RC\omega_0)^2] = 0; \quad Im^{*}: \text{imaginary part}$$

$$\omega_0 = 2\pi f_0; \quad f_0 = \frac{\sqrt{6}}{2\pi RC}$$

2.5.2.2. Condition for sustained oscillation

At the oscillation frequency f_0, the feedback network introduces an attenuation equal to $\beta(f_0)$: $\beta(f_0) = -(1/29)$.

It can be noted that besides a $(1/29)$ attenuation, the feedback network introduces at frequency f_0 a phase rotation equal to $180°$.

Voltage v_r returned from output to input undergoes a phase shift by an angle of π radians with respect to voltage v_s.

The role of the amplifier is therefore to ensure, besides amplification, an additional phase turning of $180°$, in order to meet the Barkhausen condition on the phase.

$$|A.\beta| = 1 \Leftrightarrow A.\beta(\omega = \omega 0) = 1$$

Therefore, the amplifier must bring a gain A so that $A = -29$.

$$A = -29 \Leftrightarrow |A| = 29 \text{ and } Arg(A) = \pi$$

The oscillator circuit can then be schematically represented in Figure 2.12.

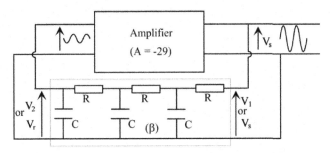

Figure 2.12. *Oscillator circuit with low-pass cells*

2.5.3. *High-pass cells*

The series impedance comprises a capacitance, and the parallel impedance comprises a resistance:

$$Z_p = R \text{ and } Z_s = (1/jC\omega)$$

2.5.3.1. *Oscillation frequency*

It is worth recalling that the amplifier gain is assumed real. In order to find the oscillation frequency of the circuit, it suffices to cancel out the imaginary part of the transfer function (β) of the feedback circuit.

$$\beta = \frac{V_r}{V_s} = \frac{[RC\omega]^3}{[RC\omega]^3 - 5[RC\omega] - 6j[RC\omega]^2 + j}$$

If $\omega = \omega_0$, then: $\text{Im}(\beta) = 0 \Leftrightarrow [1 - 6.(RC\omega_0)^2] = 0$

The expression of the oscillation frequency f_0 of the oscillator is subsequently determined:

$$f_0 = \frac{\omega_0}{2\pi} = \frac{1}{2\pi RC\sqrt{6}}$$

2.5.3.2. *Condition for sustained oscillation*

In order to find the condition required for sustained oscillation, the following is written:

$$A\beta(\omega = \omega_0) = 1; \ \beta(\omega = \omega_0) = -(1/29) \Rightarrow A = -29$$

This type of oscillator can be schematically represented by the circuit shown in Figure 2.13.

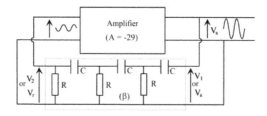

Figure 2.13. *Phase-shift oscillator circuit with high-pass cells*

2.5.4. *Phase-shift oscillator with operational amplifier*

An operational amplifier can be proposed as an amplifier element in order to obtain a phase-shift oscillator. An example that uses low-pass cells as feedback elements is schematically presented in Figure 2.14. The operational amplifier is connected in closed loop. This offers the possibility to fix the adequate gain in order to meet the oscillation condition. The operational amplifier must introduce a phase inversion and a gain equal to 29. The input resistance of the amplifier must not disturb the feedback circuit. The choice $R_1 \gg (1/C\omega_0)$ must be made for this purpose. The case of a phase-shift network with high-pass RC cells is schematically presented in Figure 2.15. The calculation of oscillator parameters is similar to that previously presented. The result for two types of oscillators will be given here.

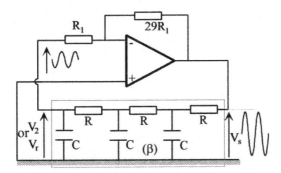

Figure 2.14. *Phase-shift oscillator (low-pass cells)*
and operational amplifier

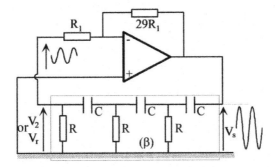

Figure 2.15. *Phase-shift oscillator (high-pass cells)*
and operational amplifier

2.5.4.1. Oscillation frequency

– Case of phase-shift network with low-pass RC cells: $f_0 = \dfrac{\omega_0}{2\pi} = \dfrac{\sqrt{6}}{2\pi RC}$

– Case of phase-shift network with high-pass RC cells: $f_0 = \dfrac{\omega_0}{2\pi} = \dfrac{1}{2\pi RC\sqrt{6}}$

2.5.4.2. Condition for sustained oscillation

According to the Barkhausen condition: $A\beta(\omega = \omega_0) = 1$; $\beta(\omega = \omega_0) = -(1/29)$. For the two types of oscillators (phase-shift network with high-pass cells or low-pass cells), the amplifier gain must be equal to $A = -29$:

In order to avoid the problem of the limited value of R_1, an operational amplifier-based follower circuit can be inserted between the feedback circuit and the amplifier.

2.5.5. RC oscillators with transistors

2.5.5.1. Generalities

It is worth seeing how phase-shift oscillators can be designed when the active element is a discrete component, such as a bipolar transistor. Such a circuit that uses a series of high-pass cells as phase-shift network is shown in Figure 2.16.

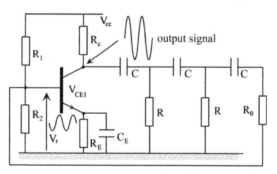

Figure 2.16. RC oscillator with bipolar transistor

The feedback network is formed by RC cells. It introduces a phase shift of 180°. The transistor is connected as a common emitter. It consequently introduces a second phase rotation of 180°. The overall phase shift of the loop that realizes the oscillator is zero. This arrangement is required by the Barkhausen condition. The elements of the oscillator diagram are the following:

– C_E: decoupling capacitance, which is equivalent to a short circuit at oscillation frequency (eliminates the effect of R_E in dynamic state)

– R_1; R_2; R_E and R_C: polarization resistances; $R_0 + h_{11} = R$

The calculation procedure is similar to the one employed in the previous cases. Therefore, only certain additional aspects are considered here. In fact, when the active element is a bipolar transistor, it is worth taking into account the polarization resistances and especially the limited value of the input resistance and the non-zero value of the output resistance of the amplifier element.

2.5.5.2. Calculation of the oscillator

2.5.5.2.1. Determination of the parameters related to the Barkhausen condition

In order to write the equation of the oscillator circuit and to determine its oscillation frequency, as well as the condition for sustained oscillation, it is important to use the equivalent diagram of the low-frequency transistor.

To reduce the calculation without introducing sensitive errors in the final result, hybrid parameters h_{12} and h_{22} are considered negligible. Only the simplified equivalent diagram of the transistor is taken into account (see Figure 2.17).

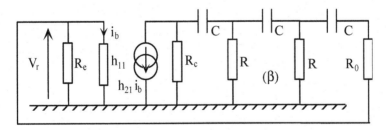

Figure 2.17. *Equivalent diagram of the RC oscillator with transistor*

Let us consider that the values of polarization resistances R_1 and R_2 are high compared to the value of input resistance h_{11} of the transistor.

$$R_e = R_1 / / R_2 = \frac{R_1 R_2}{R_1 + R_2}$$

Resistances R_1 and R_2 are chosen so that $R_e \gg h_{11}$

In the figure, v_r is the voltage returned from output to input.

The equivalent diagram of the oscillator can be transposed in a new form, as shown in Figure 2.18 (the effect of R_e is considered negligible).

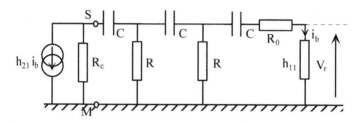

Figure 2.18. *Equivalent simplified diagram of the RC oscillator with transistors*

The voltage returned from output to input through the feedback circuit is expressed as $v_r = h_{11} \cdot i_b$

The Thévenin generator between points S and M (see Figures 2.19(a) and 2.19(b)) is characterized by: $R_{TH} = R_c$ and $V_{TH} = -R_c \cdot h_{21} \cdot i_b = v_s$

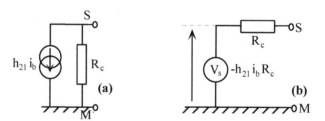

Figure 2.19. *a) Circuit seen between points S and M; b) equivalent Thévenin generator*

If the circuit seen between points S and M is replaced by the equivalent Thévenin generator in the simplified overall diagram of the oscillator, then the device shown in Figure 2.20 is obtained. This will be the basis for calculating the definition parameters of the oscillator.

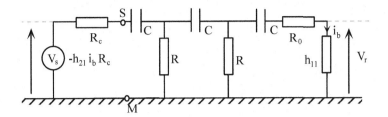

Figure 2.20. *Basic diagram of the RC oscillator with transistor*

To operate as an oscillator, the circuit must imperatively meet the Barkhausen conditions:

$$A\beta = \frac{v_s}{v_e}\frac{v_r}{v_s}$$

where voltage v_e at the amplifier input is equal to voltage v_r returned from the output to the input by the feedback phase-shift network.

$$v_e = v_r;\ v_r = h_{11}i_b \text{ and } v_s = -h_{21}R_c i_b$$

$$A = \frac{v_s}{v_e} = \frac{-h_{21}R_c}{h_{11}}$$

According to the previous relation, it can be noted that the amplifier with a transistor introduces a phase inversion.

This is perfectly normal, since the transistor is connected as a common emitter.

At oscillation frequency, the feedback circuit must therefore also introduce a second phase inversion in order to meet the Barkhausen condition on the argument.

The expression $\beta = \dfrac{v_r}{v_s}$ is determined using the circuit shown in Figure 2.21.

$$R_0 + h_{11} = R \text{ and } Z_s = (1/jC\omega)$$

Mesh 1: $v_s = (Z_s + R_c).i_1 + R(i_1 - i_2)$

Mesh 2: $0 = R.(i_2 - i_1) + Z_s.i_2 + R(i_2 - i_3)$

Mesh 3: $0 = R.(i_3 - i_2) + (Z_s + R)i_3$

$$i_3 = \frac{V}{R_0 + h_{11}}, \qquad v_r = h_{11}i_3$$

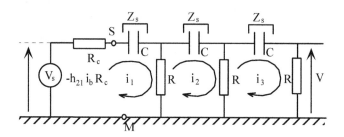

Figure 2.21. *Schematic diagram for the calculation of β*

This leads to:

$$v_r = \frac{Vh_{11}}{h_{11} + R_0}$$

The expression of v_s as a function of current i_3 can be deduced from the equations for meshes as follows:

$$v_s = i_3 \left[\left[\frac{R_c}{R} + \frac{Z_s}{R} + 1 \right] \left[4Z_s + 3R + \frac{(Z_s)^2}{R} \right] - Z_s - 2R \right]$$

If current i_3 and impedance Z_s are replaced by the equivalent expressions, then the relation between v_r and v_s is obtained.

Therefore, the expression of the transfer function of the feedback circuit is obtained:

$$\beta = \frac{v_r}{v_s} = \frac{h_{11}}{3R_c + R - \dfrac{R_C}{[RC\omega]^2} - \dfrac{5R}{[RC\omega]^2} - \dfrac{j}{[C\omega]}\left[\dfrac{4R_C}{R} - \dfrac{1}{[RC\omega]^2} + 6 \right]}$$

2.5.5.2.2. Oscillation frequency

Using the condition on the argument, the oscillation frequency f_0 is obtained:

$$Arg(A) + Arg(\beta) = 0$$

For this condition to be verified, the imaginary part of the product $A.\beta$ must be zero. A is real. Therefore, the annulation of the imaginary part of β is sufficient to deduce the oscillation frequency ($f_0 = \omega_0/2\pi$) of the circuit:

$$Im(\beta)=0 \quad \Leftrightarrow \quad \left[\frac{4R_c}{R} - \frac{1}{[RC\omega_0]^2} + 6 \right] = 0$$

$$f_0 = \frac{\omega_0}{2\pi} = \frac{1}{2\pi RC\sqrt{\left[\dfrac{4R_c}{R} + 6 \right]}}$$

2.5.5.2.3. Condition for sustained oscillation

At oscillation frequency: $A.\beta(\omega = \omega_0) = 1$

$$\beta|_{\omega=\omega_0} = \frac{v_r}{v_s} = \frac{h_{11}}{3R_c + R - \dfrac{R_c}{[RC\omega_0]^2} - \dfrac{5R}{[RC\omega_0]^2}}$$

The relation that defines the oscillation pulsation has already been established above; therefore:

$$\beta|_{\omega=\omega_0} = \frac{v_r}{v_s} = \frac{-\dfrac{h_{11}}{R_c}}{23 + \dfrac{29R}{R_c} + \dfrac{4R_c}{R}}$$

The negative sign indicates that the feedback phase-shift network introduces a phase shift equal to π. As already noted, this phase shift is complementary to that introduced by the transistor connected as a common emitter. The condition $A.\beta(\omega = \omega_0) = 1$ can this way be met:

$$A\beta|_{\omega=\omega_0} = \frac{h_{21}}{23 + \dfrac{29R}{R_c} + \dfrac{4R_c}{R}} = 1$$

Finally, this leads to the relation that ensures sustained oscillation:

$$h_{21} = 23 + \frac{29R}{R_c} + \frac{4R_c}{R}$$

The minimum gain required for obtaining an oscillation at the output of the oscillator with a transistor can be defined as follows: h_{21} is minimum when the sum "$29(R/R_c) + 4(R_c/R)$" is minimal. As it can be readily noted, the expression "$29(R/R_c) + 4(R_c/R)$" is minimal when the two component terms are equal. Indeed, this relation is the sum of two terms whose product is constant and equal to 116. This is evidenced by the direct calculation of the conditions for which the expression "$29(R/R_c) + 4(R_c/R)$" is minimal. For that purpose, let us have: $(h_{21} - 23) = y$ and $(R/R_c) = x$. Hence, $y = 29x + \dfrac{4}{x}$

$$x \to 0: y \to \infty \; ; \quad x \to \infty: y \to \infty$$

The function "y" has a minimum when its derivative is zero. The calculation of the derivative of "y" yields $\dfrac{dy}{dx} = 29 - \dfrac{4}{x^2}$

The function "y" is minimal for:

$$x = \sqrt{\frac{4}{29}}\; ; \quad \frac{R}{R_c} = \sqrt{\frac{4}{29}}\; ; \quad \frac{R_c}{R} = 2.7\; ; \quad \frac{R}{R_c} = 0.37$$

Hence, the minimum gain for which the oscillation is sustained:

$$(h_{21})\text{min} = 44.5$$

NOTES.–

1) It is worth noting that the oscillation frequency found when the active element is an operational amplifier differs from that obtained when the amplifier circuit is centered on a bipolar transistor (this obviously supposes that the RC phase-shift circuit that fixes *a priori* the oscillation frequency is identical in both cases).

2) This difference stems from the fact that the output resistance of an operational amplifier is zero, which is not the case when the amplifier element is a bipolar transistor.

3) The output resistance of the amplifier with transistor is equal to R_C. If R_C is zero, then the same oscillation frequency should be found in both cases (amplification with operational amplifier or amplification with transistor). This is certainly true. Then, the question is: why not render everything, including calculations, simpler and consider $R_C = 0$, which would yield uniformity between the operational amplifier and the bipolar transistor in the generation of sinusoidal signals?

4) Unfortunately, setting $R_C = 0$ and analyzing the condition for sustained oscillation if the amplification element is a transistor led to noting that the gain in transistor current must be infinite, which is impractical. If $R_C = 0$, the oscillation will never emerge.

5) A further difference between an operational amplifier and a bipolar transistor is the input resistance h_{11} of the latter. To eliminate its effect, it should be integrated in the last "RC" cell that constitutes the feedback phase-shift network. This has been done for the studied case.

2.6. Bridge oscillator

2.6.1. *Principle*

This type of oscillator involves two feedback loops instead of one, as shown in Figure 2.22.

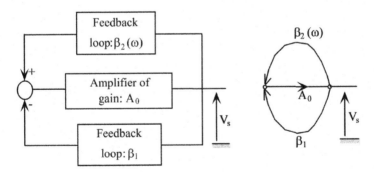

Figure 2.22. *Principle of a bridge oscillator*

The feedback loops formed by $\beta_2(\omega)$ and β_1 constitute the four branches of the bridge that, together with an amplifier of very high gain A_0, forms the oscillator to be studied here.

Here, A_0 is the amplifier gain; β_1 is the negative feedback loop and β_2 is the positive feedback loop.

One of the two feedback loops depends on frequency. For example, $\beta_2(\omega)$. The equation that governs the operation of the oscillator is:

$$A_0.[\beta_2(\omega) - \beta_1] = 1$$

At oscillation frequency f_0 (fixed by β_2), $\beta_2(\omega_0)$ is real. Certainly, the gain A_0 and the transfer function β_1 are supposed purely real (independent of the frequency parameter).

2.6.2. Principle of the bridge oscillator with operational amplifier

When the operational amplifier is considered an amplification element, the structure of the bridge oscillator in a general context is schematically presented in Figure 2.23.

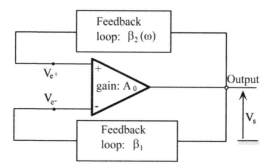

Figure 2.23. *Principle of a bridge oscillator whose active element is an operational amplifier*

According to Figure 2.23, the following relations can be established:

$$v_{e+} = \beta_2(\omega).v_s \quad \text{and} \quad v_{e-} = \beta_1.v_s \quad \Rightarrow \quad v_s = A_0.(v_{e+} - v_{e-})$$

where A_0 is the open-loop gain of the operational amplifier. If the latter is considered ideal, then A_0 is infinite.

$$\frac{v_s}{A_0} = [\beta_2(\omega) - \beta_1] v_s$$

Hence, the final relation is $A_0 [\beta_2(\omega) - \beta_1] = 1$

In the presence of a normally operating oscillator circuit, the output voltage v_s at oscillation frequency f_0 ($f_0 = \omega_0/2\pi$) is not zero. The gain A_0 of the operational amplifier is very high and it can be considered infinite in the ideal case. Under these conditions:

$$A_0 \longrightarrow \infty ; \quad \Rightarrow \quad \beta_2(\omega_0) \longrightarrow \beta_1$$

Thus, at the oscillation frequency, the difference between β_2 and β_1 is practically zero.

2.6.3. Study of the bridge oscillator in the general case

Let us consider the circuit shown in Figure 2.24. The nature of impedances Z_1, Z_2, Z_3 and Z_4 that form the branches of the bridge is not *a priori* defined. The condition for circuit oscillation is given by:

$$A_0.(\beta_2(\omega) - \beta_1) = 1$$

$$A_0 = \frac{v_s}{v_{e^+} - v_{e^-}} = \frac{V_s}{V_1}$$

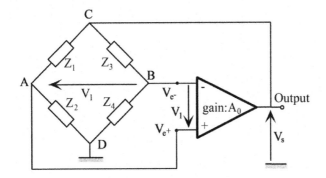

Figure 2.24. *Bridge oscillator in the most general case*

For the sake of simplicity, let us have $\beta = (\beta_2(\omega) - \beta_1)$,

$$A_0\beta = \frac{v_s}{V_1} \cdot \frac{V_1}{v_s}$$

The function β can be expressed by the following relation:

$$\beta = \frac{V_1}{V_s} = \frac{V_A - V_B}{V_C - V_D} = \frac{V_A - V_D}{V_C - V_D} + \frac{V_D - V_B}{V_C - V_D}$$

$$\beta = \frac{V_{AD}}{V_{CD}} - \frac{V_{BD}}{V_{CD}}$$

$$V_{AD} = \frac{Z_2}{Z_1 + Z_2} V_{CD} \quad \text{and} \quad V_{BD} = \frac{Z_4}{Z_3 + Z_4} V_{CD}$$

Finally, for any circuit represented by a bridge, such as the one shown in Figure 2.24, the general expression of β is obtained:

$$\beta = [\beta_2(\omega) - \beta_1] = \frac{1}{1 + \frac{Z_1}{Z_2}} - \frac{1}{1 + \frac{Z_3}{Z_4}}$$

2.6.4. *Study of Wien bridge oscillator*

One of the most commonly employed oscillators at low frequency is the one that uses the Wien bridge (see Figure 2.25) as a feedback circuit.

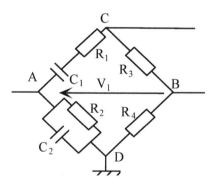

Figure 2.25. *Representative circuit of Wien bridge*

Impedances Z_1, Z_2, Z_3 and Z_4 that form the various branches of this bridge are defined by the following relations:

$$Z_1 = R_1 + \frac{1}{jC_1\omega}; \quad Z_2 = \frac{R_2}{1 + jR_2C_2\omega}; \quad Z_3 = R_3; \quad Z_4 = R_4$$

The relation that defines β is:

$$\beta = [\beta_2(\omega) - \beta_1] = \frac{1}{1 + \dfrac{[R_1 + (1/jC_1\omega)][1 + jR_2C_2\omega]}{R_2}} - \frac{1}{1 + \dfrac{R_3}{R_4}}$$

where $\beta_2(\omega)$ and β_1 are readily identified.

When the aforementioned bridge is properly connected to an operational amplifier, a Wien bridge oscillator is obtained (see Figure 2.26).

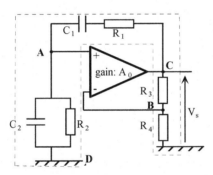

Figure 2.26. *Wien bridge oscillator*

Having presented the diagram of the bridge oscillator in the general case and the Wien bridge oscillator in a specific case, the next step is to calculate the elements of this oscillator. For this purpose, the relation that defines the oscillation frequency must be established, as well as the condition for sustained oscillation.

2.6.4.1. Oscillation frequency of the Wien bridge oscillator

To calculate the oscillation frequency, the oscillator definition equation must be employed: $A_0 . \beta = A_0 . (\beta_2(\omega_0) - \beta_1) = 1$

$$\beta_2(\omega) - \beta_1 = \frac{1}{A_0} \longrightarrow 0$$

The above can be summarized by stating that the imaginary and real parts of the difference $(\beta_2(\omega 0) - \beta_1)$ must be zero.

The annulation of the imaginary part of this difference allows determination of the oscillation frequency, and the annulation of the real part yields the condition for sustained oscillation.

$$\mathrm{Im}(\beta) = \mathrm{Im}\ [\ \beta_2(\omega_0) - \beta_1] = \frac{1}{jC_1\omega_0} + jR_1R_2C_2\omega_0 = 0$$

Hence, the oscillation frequency is:

$$f_0 = \frac{1}{2\pi\sqrt{R_1R_2C_1C_2}}$$

If $R_1 = R_2 = R$ and $C_1 = C_2 = C$ (very frequent case), the oscillation frequency is defined by the following relation:

$$f_0 = \frac{1}{2\pi RC}$$

2.6.4.2. Condition for sustained oscillation of the Wien bridge oscillator

In order to find the condition for sustained oscillation, the real part of β (Re (β)) is annulled.

$$\mathrm{Re}\ (\beta) = \mathrm{Re}\ [\beta_2(\omega_0) - \beta_1] = 0$$

$$\mathrm{Re}(\beta) = \frac{1}{1 + \dfrac{R_1}{R_2} + \dfrac{C_2}{C_1}} - \frac{1}{1 + \dfrac{R_3}{R_4}} = 0 \quad \text{Þ} \quad \frac{R_1}{R_2} + \frac{C_2}{C_1} = \frac{R_3}{R_4}$$

If $R_1 = R_2 = R$ and $C_1 = C_2 = C$, then the condition for sustained oscillation comes down to the following relation:

$$R_3 = 2R_4$$

2.6.5. *Study of the Wien bridge oscillator as one feedback branch oscillator*

As already noted, bridge oscillators have two feedback branches. The previous study has evidenced a certain calculation approach to this type of oscillator.

This type of circuit can nevertheless be dealt with as a classic oscillator, with only one feedback branch and an amplifier circuit with purely real and well-defined gain A_1, as shown in Figure 2.27.

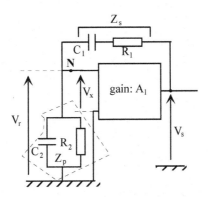

Figure 2.27. *Wien bridge oscillator treated as one feedback branch oscillator*

For the sake of simplicity, let us consider that the input impedance of the amplifier is very high and does in no way charge the feedback circuit.

There is sustained oscillation provided that relation $A_1.\beta = 1$ is verified, where β is the transfer function of the feedback circuit. Since the input impedance of the amplifier has been supposed high, a cutoff at point "N" can be done, followed by the calculation of β.

$$\beta = \frac{v_r}{v_s} \qquad \beta = \frac{Z_p}{Z_s + Z_p} = \frac{1}{1 + \dfrac{Z_s}{Z_p}}$$

$$Z_s = R_1 + \frac{1}{jC_1\omega}; \qquad Z_p = \frac{R_2}{1 + jR_2C_2\omega}$$

$$\beta = \frac{1}{1 + \dfrac{R_1}{R_2} + \dfrac{C_2}{C_1} + j(R_1C_2\omega - \dfrac{1}{R_2C_1\omega})}$$

The amplifier gain A_1 is assumed purely real (which is generally the case for all types of low frequency amplifiers). Under these conditions, the transfer function β should also be purely real.

2.6.5.1. Oscillation frequency

In order to calculate the oscillation frequency, it suffices to annul the imaginary part related to the transfer function β of the feedback circuit.

$$\text{Imaginary } (\beta) = 0 \quad \Leftrightarrow \quad (R_1 C_2 \omega_0 - \frac{1}{R_2 C_1 \omega_0}) = 0$$

Hence, the oscillation frequency is:

$$f_0 = \frac{1}{2\pi\sqrt{R_1 R_2 C_1 C_2}}$$

If $R_1 = R_2 = R$ and $C_1 = C_2 = C$, then $f_0 = \frac{1}{2\pi RC}$

2.6.5.2. Condition for sustained oscillation

For this purpose: $A_1 . \beta \ (\omega = \omega_0) = 1$

$$A_1 = 1 + \frac{R_1}{R_2} + \frac{C_2}{C_1}$$

If $R_1 = R_2 = R$ and $C_1 = C_2 = C$, then $\mathbf{A_1 = 3}$

Thus, the existence of sustained oscillation requires a closed-loop gain of the amplifier equal to 3. This compensates the attenuation introduced by the feedback circuit. At oscillation frequency, the transfer function of the feedback circuit is positive, and consequently, there is no phase shift between the output voltage and the voltage returned to the amplifier input. Therefore, the latter should introduce no phase shift in order to meet the oscillation conditions of the circuit. Thus, a type of amplifier can be designed that meets the already mentioned specifications using, for example, an operational amplifier connected in closed loop, as indicated in Figure 2.28. It is worth noting that the voltage at the input of amplifier of gain A_1 is equal to the voltage returned from the output to input.

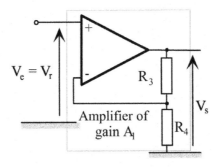

Figure 2.28. *Amplifier of gain A₁ for the Wien bridge oscillator*

The calculation of gain A_1 introduced by this amplifier is given by:

$A_1 = (R_3 + R_4)/R_4$; $A_1 = 3$ when $R_1 = R_2 = R$ and $C_1 = C_2 = C$. The condition for sustained oscillation imposes $R_3 = 2.R_4$

2.7. Band-pass filter oscillator

2.7.1. *Feedback circuit*

Let us consider the cascade of two RC cells, the first being connected as a high-pass filter, while the second plays the role of a low-pass filter of the first order, as schematically represented by the circuit shown in Figure 2.29.

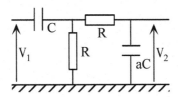

Figure 2.29. *High-pass and low-pass cells assembled in cascade to obtain a band-pass filter*

The transfer function of this circuit is expressed as:

$$\frac{v_2}{v_1} = \beta(\omega); \qquad \beta(\omega) = \frac{RC\omega}{RC\omega(1+2a) - j(1 - a(RC\omega)^2)}$$

where "a" is a real positive parameter.

The overall frequency response of the filter thus realized is represented in Figure 2.30.

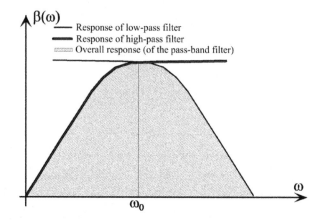

Figure 2.30. *Frequency response of the band-pass filter realized by cascading a high-pass filter and a low-pass filter*

It can be noted that this circuit favors the passage of a certain frequency f_0 (with $f_0 = \omega_0/2\pi$) with respect to any other frequency. It can therefore be used as a feedback element of an oscillator operating at frequency f_0.

2.7.2. Oscillator circuit

Based on the previously presented feedback circuit, an example of an oscillator with an operational amplifier is schematically shown in Figure 2.31. The amplifier element has a high input impedance. It does not charge the feedback network.

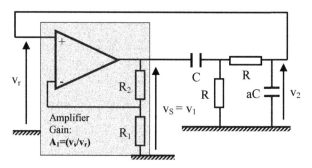

Figure 2.31. *Oscillator with elementary band-pass filter*

After the presentation of the oscillator, the next step is the calculation of the oscillation frequency and the condition for sustained oscillation.

2.7.3. *Oscillation frequency*

In order to find the oscillation frequency of the circuit shown in Figure 2.31, the Barkhausen condition on the argument should be verified:

$$\text{Arg}(A_1) + \text{Arg}(\beta) = 0 \ [2\pi]$$

The closed-loop gain A_1 of the amplifier is given by:

$$A_1 = \frac{v_s}{v_r} = \frac{R_2 + R_1}{R_1}$$

The gain A_1 is positive real. The amplifier introduces no phase shift between output and input $[\text{Arg}(A_1) = 0]$. It can therefore be stated that the feedback circuit should introduce no phase shift between its input and output at oscillation frequency f_0 $(f_0 = \omega_0/2\pi)$.

Otherwise, the circuit cannot operate as an oscillator. The transfer function of the feedback circuit has already been calculated. It is worth recalling its expression:

$$\frac{v_2}{v_1} = \beta(\omega); \qquad \beta(\omega) = \frac{RC\omega}{RC\omega(1+2a) - j(1 - a(RC\omega)^2)}$$

The oscillation frequency f_0 is obtained as follows:

$$\text{Arg}(\beta) = 0 \Leftrightarrow \text{imaginary part } (\beta) = 0;$$

$$\text{Im}(\beta) = 0 \Leftrightarrow (1 - a(RC\omega_0)^2) = 0$$

Hence, the oscillation frequency of the circuit is:

$$f_0 = \frac{\omega_0}{2\pi} = \frac{1}{2\pi RC\sqrt{a}}$$

2.7.4. Condition for sustained oscillation

The condition for sustained oscillation applies to the expression represented by the product $A_1.\beta(\omega_0)$. This product should verify the following equality at oscillation frequency: $A_1.\beta(\omega = \omega_0) = 1$

The calculation of the expression of the transfer function of the feedback circuit at oscillation frequency yields:

$$\beta(\omega = \omega_0) = \frac{1}{(1+2a)} \qquad A_1.\beta(\omega = \omega_0) = \frac{R_2+R_1}{R_1}\frac{1}{(1+2a)} = 1$$

Finally, this leads to the condition to be met for sustained oscillations:

$$a = \frac{R_2}{2R_1}$$

2.8. Generator of sinusoidal waves with shaper

2.8.1. Principle

This type of circuit has no direct relationship with feedback oscillators. Nevertheless, it allows the generation of sinusoidal waves in the low-frequency domain. The principle of this device is schematically represented by the block diagram shown in Figure 2.32.

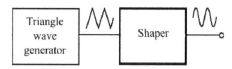

Figure 2.32. *Principle of sinusoidal wave generator with shaper*

The sinusoidal wave generator with shaper is limited in terms of spectral range. Indeed, it can only generate signals, the frequency of which rarely exceeds the megahertz range. The sinusoidal wave is reconstructed by approximations from a triangle wave.

This role is ensured by the shaper circuit. The positive and negative slopes of the triangle wave must have the same amplitude.

2.8.2. *Generation of the triangle wave*

2.8.2.1. *Definition*

There are many circuits that can generate a triangle wave. For the case in point, and in order to facilitate knowledge acquisition and comprehension, a simple device elaborated on the basis of classic circuits, such as operational amplifiers, will be studied (see Figure 2.33).

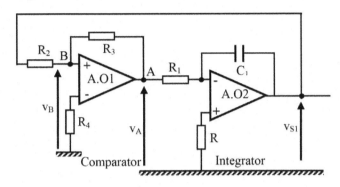

Figure 2.33. *Generator of triangle waves with operational amplifiers*

This circuit is in fact formed of a comparator, the reference voltage (threshold voltage) of which is fixed at 0 V, and an integrating circuit.

2.8.2.2. *Operation*

The comparator output only takes one of two possible states: "high state or low state". In order to explain the circuit operation, voltage v_A is assumed at its high state (assuming the opposite is also possible, in order to reach the same result). Operational amplifiers are supplied at $\pm V_{cc}$. It can be written with approximation $v_A = +V_{CC}$.

High or low saturation of operational amplifiers rarely reaches the value $\pm V_{CC}$ because of a waste voltage. As a first step, output voltage v_{s1} can be expressed by the following relation:

$$v_{s1} = \frac{-1}{R_1 C_1} \int_0^t V_A \, dt \Rightarrow v_{s1} = \frac{-1}{R_1 C_1} V_{cc}.t$$

Each operational amplifier is considered ideal. Resistances R_2 and R_3 are crossed by the same current. It can then be shown that:

$$v_B = \frac{R_2 v_A + R_3 v_{s1}}{R_2 + R_3}$$

If $R_2 = R_3$ (for the sake of simplicity, only this case of equality will be considered in the following): $v_B = \dfrac{v_A + v_{s1}}{2}$

As already mentioned, the threshold of the comparator is fixed at zero. The shift of the state of its output depends on the direction of passage through zero of voltage v_B applied at the input of this comparator.

$$v_A = +V_{cc}; \quad v_{s1} = \frac{-1}{R_1 C_1} V_{cc}.t; \quad v_B = \frac{\dfrac{-1}{R_1 C_1} V_{cc}.t + V_{cc}}{2}$$

Voltage v_B evolves in time according to a negative slope, starting from value $(V_{CC}/2)$.

At $t = \tau = R_1.C_1$, voltage v_B reaches and drops below zero (v_B becomes slightly < 0). The output state of the comparator shifts from the high state to the low state.

Moreover, at this precise instant, the value of voltage v_B goes to $-V_{CC}$ due to the abrupt passage of v_A from $+V_{CC}$ to $-V_{CC}$.

Starting from this instant, which has been chosen as the origin of time for the steady state, voltage v_B will increase linearly in time with a positive slope (initial voltage is equal to $-V_{cc}$).

$$v_A = -V_{cc}; \quad v_{s1} = \frac{1}{R_1 C_1} V_{cc}.t - V_{cc}; \quad v_B = \frac{\dfrac{1}{R_1 C_1} V_{cc}.t - 2V_{cc}}{2}$$

When voltage v_B reaches and slightly exceeds zero, in the increasing direction, the output of the comparator shifts from low state to high state.

Voltages v_A, v_{S1} and v_B take new values, and the previously described cycle is repeated indefinitely.

$$v_A = +V_{cc}; v_{s1} = \frac{-1}{R_1 C_1} V_{cc}.t + V_{cc};$$

$$v_B = \frac{\dfrac{-1}{R_1 C_1} V_{cc}.t + 2V_{cc}}{2}$$

Signal v_{s1} at generator output has a triangular form. It is effectively ready for being processed by the shaper in order to obtain the desired sinusoidal wave.

The various signals employed at the level of the generator of the triangle wave are given in Figure 2.34.

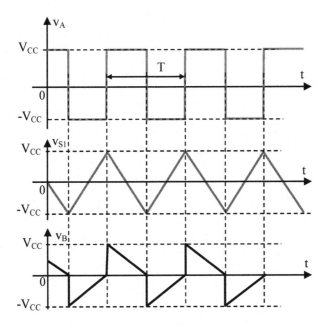

Figure 2.34. *Signals employed at the level of the generator of the triangle wave. For a color version of this figure, see www.iste.co.uk/haraoubia/nonlinear1.zip*

2.8.3. *Shaper circuit*

The role of the shaper is to reconstruct by approximations, starting from a triangle wave, a practically pure sinusoidal wave. The electric circuit representative for a type of elementary shaper with diodes with three thresholds is shown in Figure 2.35.

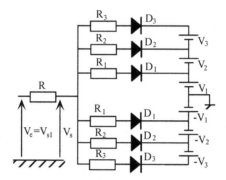

Figure 2.35. *Elementary shaper with diodes*

2.8.4. *Shaper operation*

For a simpler explanation of the operation, all the diodes that constitute the shaper circuit are assumed ideal. The diagram in Figure 2.36 explains the operation of this device. The circuit involves a certain number of symmetrical levels of positive and negative voltages (V_1, V_2, V_3 and $-V_1$, $-V_2$, $-V_3$), which are used as references. The charge of the shaper must be high with respect to resistance R.

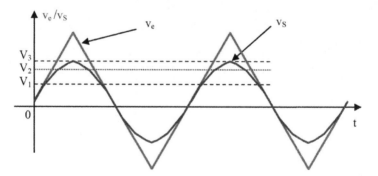

Figure 2.36. *Sinusoidal wave emerging from a triangle signal. For a color version of this figure, see www.iste.co.uk/haraoubia/nonlinear1.zip*

Only the positive half-cycle will be studied. The negative half-cycle can be treated similarly.

The shaper circuit operates as follows:

$$v_e < V_1 : v_s = v_e$$

When $V_1 < v_e < V_2$, diode D_1 is conducting and then:

$$v_s = \frac{R_1 v_e + R V_1}{R_1 + R}$$

Between V_1 and V_2, the input voltage can be written in the following form:

$$v_e = V_1 + \Delta v_e, \text{ where } \Delta v_e \text{ is a linear variation.}$$

Therefore:

$$v_s = \frac{R_1(V_1 + \Delta v_e) + R V_1}{R_1 + R} = V_1 + \frac{R_1 \Delta v_e}{R_1 + R}$$

The variation of output voltage v_s is always a linear function of v_e, but it is slower than that of input voltage v_e due to the intervention of attenuator network $R - R_1$. Now, the input voltage v_e exceeds the value V_2 and becomes $V_2 < v_e < V_3$.

Diodes D_1 and D_2 conduct and bring in resistances R_1 and R_2. They send the input voltage toward the output, but with an attenuation coefficient that is more significant than in the previous case. The same is applicable when voltage v_e exceeds the value V_3; all the diodes are conducting and the transfer from the input voltage to the output is done with a higher attenuation due to the intervention of resistances R_1, R_2 and R_3. Thus, at each variation beyond the reference voltages, the input voltage is transmitted to the output with an increasingly weaker coefficient (increasingly stronger attenuation). This leads to the sinusoidal form that is schematically represented in Figure 2.36.

2.8.5. Frequency of the output signal

The frequency of the sinusoidal output signal is the frequency of the generated triangle signal. Therefore, it is sufficient to calculate the frequency of the latter.

The calculation of the period T of the triangle signal requires the calculation of durations T_1 and T_2 (see Figure 2.37).

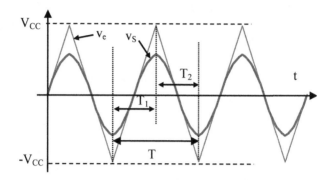

Figure 2.37. *Correspondence between the period of the triangle signal and the period of the output sinusoidal signal. For a color version of this figure, see www.iste.co.uk/haraoubia/nonlinear1.zip*

Calculating the period of the sinusoidal signal requires the calculation of T_1 and T_2.

$$T = T_1 + T_2$$

– Calculation of T_1:

For the calculation of T_1, it is sufficient to revisit the expression that defines the triangle signal v_e or v_{s1} ($v_e = v_{s1}$) applied at the input of the shaper. Let us recall that when the triangle signal evolves with a positive slope, the various existing signals have the following expressions:

$$v_A = -V_{CC} \quad \text{and} \quad v_{s1} = \frac{1}{R_1 C_1} V_{cc}.t - V_{cc}$$

The signal of interest is v_{s1}. At $t = 0$, $v_{s1} = -V_{cc}$. On the contrary, at $t = T_1$, $v_{s1} = V_{cc}$.

This allows writing at $t = T_1$:

$$v_e = v_{s1} = \frac{1}{R_1 C_1} V_{cc}.T_1 - V_{cc} = V_{cc} ; \quad T_1 = 2R_1 C_1$$

– Calculation of T_2:

The expression that defines v_{s1} when the evolution of its slope is negative is:

$$v_{s1} = \frac{-1}{R_1 C_1} V_{cc}.t + V_{cc}$$

At $t = 0$, $v_{s1} = V_{cc}$. On the contrary, at $t = T_2$, $v_{s1} = -V_{cc}$.

$$v_e = v_{s1} = \frac{-1}{R_1 C_1} V_{cc} T_2 + V_{cc} = -V_{cc} ; \quad \Rightarrow \quad T_2 = 2R_1 C_1$$

$$T = 4R_1 C_1$$

The frequency of the output sinusoidal signal is then defined as: $f = \dfrac{1}{T} = \dfrac{0.25}{R_1 C_1}$

High-frequency Oscillators

3.1. Elementary high-frequency oscillator

3.1.1. *Equation*

Let us consider the circuit in Figure 3.1. Capacitor C is first charged at voltage V_0. At instant t = 0, switch K is closed.

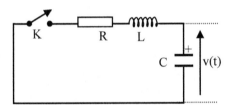

Figure 3.1. *RLC circuit, C is charged at continuous voltage V_0*

When the switch K is closed, a current "i" flows across the RLC circuit and the following equation can be written:

$$Ri + L\frac{di}{dt} + \frac{1}{C}\int_0^t idt = 0$$

By hypothesis, it is known that: $\mathbf{v(0) = V_0}$, $\dfrac{dv}{dt}(0) = 0$ and $\mathbf{i(0) = 0}$

$$v = \frac{1}{C}\int_0^t idt \Rightarrow i = C\frac{dv}{dt}$$

$$RC\frac{dv}{dt} + LC\frac{d^2v}{dt^2} + v = 0$$

Finally, the following differential equation is obtained:

$$\frac{d^2v}{dt^2} + \frac{R}{L}\frac{dv}{dt} + \frac{1}{LC}v = 0$$

It is a second-order differential equation with constant coefficients. Let us analyze and solve this equation. This involves writing:

$$\frac{d^2v}{dt^2} + \frac{R}{L}\frac{dv}{dt} + \frac{1}{LC}v = \frac{d^2v}{dt^2} + a\frac{dv}{dt} + bv = 0 \; ; \text{ with } a = \frac{R}{L}; b = \frac{1}{LC}$$

The characteristic equation of this differential equation is:

$$r^2 + ar + b = 0$$

This characteristic equation has two solutions:

$$r_1 = \frac{-a + \sqrt{a^2 - 4b}}{2} = -\frac{R}{2L} + \sqrt{\frac{R^2}{4L} - \frac{1}{LC}}$$

$$r_2 = \frac{-a - \sqrt{a^2 - 4b}}{2} = -\frac{R}{2L} - \sqrt{\frac{R^2}{4L} - \frac{1}{LC}}$$

In order to find the solution to the differential equation, several cases should be considered:

$$\frac{R^2}{4L} - \frac{1}{LC} > 0; \qquad \frac{R^2}{4L} - \frac{1}{LC} = 0 \qquad \text{and} \qquad \frac{R^2}{4L} - \frac{1}{LC} < 0$$

3.1.1.1. *First case:* $\dfrac{R^2}{4L} - \dfrac{1}{LC} > 0$

The differential equation admits the following aperiodic solution:

$$v(t) = Ae^{r_1 t} + Be^{r_2 t}$$

A and B are two constants that can be determined using boundary conditions, as follows:

$$v(0) = A+B = V_0 \text{ and } \frac{dv}{dt}(0) = 0$$

$$\frac{dv}{dt} = r_1 A e^{r_1 t} + r_2 . B e^{r_2 t}$$

$$\frac{dv(0)}{dt} = 0 = r_1 A + r_2 . B$$

Based on these equations, the relations that define constants A and B and the expression of voltage v(t) are determined.

$$A = -\frac{r_2 V_0}{r_1 - r_2} \quad \text{and} \quad B = \frac{r_1 V_0}{r_1 - r_2} ; \quad v(t) = \frac{V_0}{r_1 - r_2} \left[-r_2 e^{r_1 t} + r_1 e^{r_2 t} \right]$$

As can be recalled:

$$r_1 = -\frac{R}{2L} + \sqrt{\frac{R^2}{4L} - \frac{1}{LC}} \quad \text{and} \quad r_2 = -\frac{R}{2L} - \sqrt{\frac{R^2}{4L} - \frac{1}{LC}}$$

Since $\dfrac{R^2}{4L} > \dfrac{1}{LC}$, then necessarily: $r_1 < 0$ and $r_2 < 0$.

The representation of the evolution of voltage v(t) across capacitor C is given by the diagram in Figure 3.2.

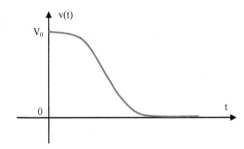

Figure 3.2. *Evolution of voltage v(t) across capacitor*

3.1.1.2. *Second case:* $\dfrac{R^2}{4L} - \dfrac{1}{LC} = 0$

Under these conditions, a critical solution is obtained, with:

$$r_1 = r_2 = r_0 = -\frac{R}{2L}$$

The differential equation admits the following solution:

$$v(t) = A.e^{r_0 t} + B.te^{r_0 t}$$

Constants A and B can be determined using boundary conditions, as follows:
$$v(0) = A = V_0$$

$$\frac{dv}{dt} = r_0 A.e^{r_0 t} + B.e^{r_0 t} + B.tr_0 e^{r_0 t}$$

Based on the first derivative of voltage across the capacitor, a second equation allowing the determination of constant B can be found:

$$\frac{dv(0)}{dt} = 0 = r_0 A + B \quad \Rightarrow \quad B = -r_0 A = -r_0 V_0$$

$$v(t) = V_0 e^{r_0 t}(1 - r_0 t)$$

It is also an aperiodic solution.

3.1.1.3. *Third case:* $\dfrac{R^2}{4L} < \dfrac{1}{LC}$

The solution to the characteristic equation of the differential equation is given by the following complex roots:

$$r_1 = -\frac{R}{2L} + j\sqrt{\frac{1}{LC} - \frac{R^2}{4L}} \quad \text{and} \quad r_2 = -\frac{R}{2L} - j\sqrt{\frac{1}{LC} - \frac{R^2}{4L}}$$

having: $r_1 = \alpha + j\beta \quad$ and $\quad r_2 = \alpha - j\beta$

with: $\alpha = -\dfrac{R}{2L} \quad$ and $\quad \beta = \sqrt{\dfrac{1}{LC} - \dfrac{R^2}{4L}}$

Let us recall that the solution to the differential equation has the following form:

$$v(t) = A.e^{r_1 t} + B.e^{r_2 t}$$

The determination of constants A and B yields:

$$v(t) = \frac{V_0}{r_1 - r_2}\left[-r_2 e^{r_1 t} + r_1 e^{r_2 t}\right]$$

After replacing r_1 and r_2 with their equivalent expression, the final relation that defines the evolution of voltage v(t) across the capacitor is obtained:

$$v(t) = \frac{V_0}{\beta} e^{\alpha t} \left[\beta\cos(\beta t) - \alpha\sin(\beta t)\right]$$

Let us recall that: $\alpha = -\dfrac{R}{2L}$ and $\beta = \sqrt{\dfrac{1}{LC} - \dfrac{R^2}{4L}}$

In the present case, α is a negative quantity and β is a positive quantity. In a theoretical context, it can be assumed, for example, that resistance R can also be negative and, consequently, parameter α can be positive. This has quite interesting consequences.

After trigonometric transformation, the following relation is obtained:

$$v(t) = \frac{V_0}{\beta} e^{\alpha t} \left[\sqrt{\alpha^2 + \beta^2}.\sin(\beta t + \phi)\right]$$

with: $\phi = Arctg(\dfrac{\beta}{\alpha});$ $\alpha > 0$ and $\phi = Arctg(\dfrac{\beta}{\alpha}) + \pi;$ $\alpha < 0$

3.1.2. *Study of the evolution of output voltage: sinusoidal condition*

In order to determine the evolution of the output voltage v(t) across the capacitor, it should be noted that resistance R (which is obviously positive) is the energy consumer element. Resistance R is a disturbing element.

As already mentioned, it can also be considered that resistance R has a negative value (negative resistances can be realized in practice). This case is also worth studying. The absence of R (R=0) should also be considered. These three cases will be studied separately and the evolution of voltage v(t) across the capacitor will be given for each of them.

3.1.2.1. *A case when resistance R is present*

When resistance R is present, parameter α is negative, and the expression that defines v(t) is:

$$v(t) = \frac{V_0}{\beta} e^{\alpha t} \left[\sqrt{\alpha^2 + \beta^2} . \sin(\beta t + \phi) \right] \quad \text{and}$$

$$\phi = \text{Arctg}(\frac{\beta}{\alpha}) + \pi; \quad \alpha < 0$$

The term $e^{\alpha t}$ represents an attenuation that increases in time. The representation of voltage v(t) across the capacitor (Figure 3.3) is a damped sinusoid. Damping depends on the value of R. The more significant the value of resistance R, the more rapid the attenuation of v(t).

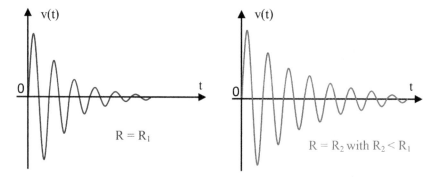

Figure 3.3. *Evolution of voltage v(t) across the capacitor when $\alpha < 0$*

3.1.2.2. *A case when negative resistance R is present*

The voltage across the capacitor verifies the following relation:

$$v(t) = \frac{V_0}{\beta} e^{\alpha t} \left[\sqrt{\alpha^2 + \beta^2} . \sin(\beta t + \phi) \right]$$

$$\phi = \text{Arctg}(\frac{\beta}{\alpha}); \quad \alpha > 0$$

The elements that have changed with respect to the previous case are the signs of parameter α and the value of the initial phase. The term $e^{\alpha t}$ represents an exponential growth of the sinusoid that represents v(t) as a function of time.

$$\frac{R^2}{4L} - \frac{1}{LC} < 0 \quad \text{and} \quad R < 0$$

It is true that resistances known and employed in everyday life are components with positive values.

Nevertheless, as already noted, there are electronic devices whose characteristics are equivalent to those of negative resistances.

For example, the characteristic of the tunnel diode features one part where the latter behaves as a negative resistance.

The representation of voltage v(t) across the capacitor is given by the diagram in Figure 3.4.

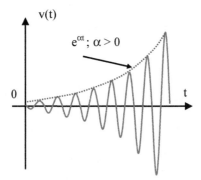

Figure 3.4. *Evolution of voltage v(t) across capacitor when α is positive*

The resulting sinusoidal signal has a nonlinear increase. Its increase rate depends on the value of negative resistance "R".

3.1.2.3. *A case when resistance R is zero*

The expression of voltage v(t) comes down to:

$$v(t) = V_0 \sin(\beta t + \phi) \quad \text{and} \quad \phi = \frac{\pi}{2}$$

$$v(t) = V_0 \cos(\beta t)$$

The energy dissipating resistance no longer exists (it is assumed that its effect has been eliminated). When the LC circuit is closed, an exchange of electric energy (across the capacitor) and magnetic energy (across the self-inductance) takes place without losses. This leads to the creation of a sustained sine wave, as shown in Figure 3.5.

$$v(t) = V_0 \cos(\frac{1}{\sqrt{LC}} t) = V_0 \cos(\omega t) = V_0 \cos(2\pi f t)$$

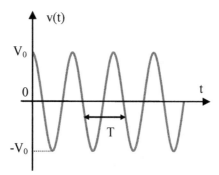

Figure 3.5. *Evolution of the signal v(t) across capacitor when R=0*

The peak amplitude of the generated signal is equal to V_0 and the oscillation frequency f_0 is equal to:

$$f_0 = \frac{1}{2\pi\sqrt{LC}}$$

It can also be found that, at instant $t = 0$, the voltage across the capacitor is effectively V_0.

Thus, if an ideal capacitor and an ideal (without losses) self-inductance were available, a high-frequency oscillator would be readily available, just by charging in the beginning the capacitor at a given voltage (here V_0) and connecting it to the self-inductance. However, ideal components do not exist. Therefore, the losses of an LC circuit (represented here by resistance R) will have to be compensated by an associated active circuit, in order to be able to realize an oscillator that delivers a sinusoidal wave of fixed amplitude.

3.2. High-frequency oscillators with discrete components

3.2.1. *Introduction*

In order to realize high-frequency oscillators and compensate the losses related to feedback circuits generally composed of passive components, discrete active components are used, such as bipolar transistors or field-effect transistors. Integrated circuits generally have limited bandwidth, which is the reason why they cannot satisfy the demands of high and very high-frequency oscillators. Integrated circuits are generally used in low-frequency circuits. There are however several integrated circuits that can reach several hundred MHz. An example is the integrated circuit of Motorola, MC 1648, whose fabrication uses ECL technology.

This section focuses on oscillator circuits with discrete components, and mainly on the bipolar transistors.

3.2.2. *Equivalent diagram of the bipolar transistor*

The equivalent diagram of a bipolar transistor in high and very high frequencies (or equivalent Giacoletto diagram) is shown in Figure 3.6. The transistor is considered connected as a common emitter.

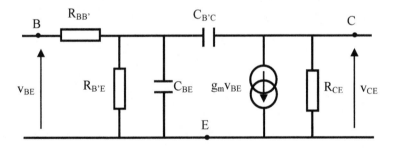

Figure 3.6. *Equivalent diagram of a bipolar transistor in high and very high frequency*

Capacitance $C_{B'E}$ is a diffusion capacitance. It is equivalent to the diffusion capacitance of a diode with forward polarized junction. The element $C_{B'C}$ represents the transition capacitance. It is similar to the transition capacitance of a reverse polarized PN junction. Its value is generally below a hundred picofarads (10^{-12}F). The element $R_{BB'}$ represents the resistance between the base and point B', which is at the limit of the emitter's junction. There are also the other elements that are normally found in the weak signal and low-frequency models: $R_{B'E} = h_{11e}$; $R_{CE} = 1/h_{22e}$ and $g_m = h_{21e}/h_{11e}$.

In general, the frequency provided by datasheets is the transition frequency f_T. The reason is very simple. The measurement of this frequency involves no great difficulty. At this frequency f_T, the current gain (h_{21}) is equal to 1. The Giacoletto diagram is valid only at working frequency f.

With: $f < 0.5 f_T$

When operating at frequencies much lower than the transition frequency, the equivalent low-frequency model can be used as an equivalent diagram for the bipolar transistor, since the influence of elements ($C_{B'E}$, $C_{B'C}$ $R_{BB'}$) that are represented in the Giacoletto diagram remains negligible.

In order to avoid rendering the study of high- and very high-frequency oscillators too complicated, the present case involves bipolar transistors that have very high transition frequency compared to the working frequency of the oscillator.

Under these conditions, the following approximations can be made: $R_{BB'} = 0$, capacitors $C_{B'E}$ and $C_{B'C}$ have very high impedances at working frequency, and they can be assimilated to open circuits with respect to the resistive impedances employed (internal to the transistor and polarization resistances). The equivalent diagram of the transistor comes down to the circuit shown in Figure 3.7.

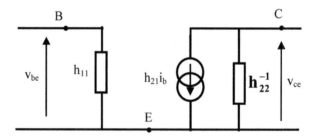

Figure 3.7. *Simplified equivalent diagram of a bipolar transistor when transition frequency is very high compared to working frequency*

3.3. Study of oscillators with bipolar transistors

3.3.1. *Operating equation*

The input impedance of a transistor has a limited value, which is generally low. This impedance may present the risk of loading the feedback circuit. Under these conditions, it is not possible to study the amplification circuit separately from the feedback circuit in the absence of impedance balance.

In what follows, another methodology (different from that proposed in chapter 2) will be proposed for the study of oscillator circuits, one that can take into account the problems related to the input impedance and even to complex parameters of the amplification circuit, if necessary (when the transition frequency is not very high compared with the oscillation or working frequency). The general design of the oscillator circuit is represented by the structure schematically shown in Figure 3.8.

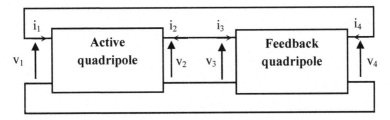

Figure 3.8. *General structure of an oscillator*

Proper operation of the oscillator requires the existence of v_2 and v_4.

$v_2 \neq 0$ and $v_4 \neq 0$

When operating at high and very high frequency, the bipolar transistor (active element) is defined by admittances Y_{ij}.

The following can be written for the active quadripole:

$$i_1 = Y_{11}v_1 + Y_{12}v_2$$
$$i_2 = Y_{21}v_1 + Y_{22}v_2$$

The passive feedback quadripole can be defined either by its impedance parameters or by its admittance parameters. A choice has been made to define it by its impedance parameters.

$$v_3 = Z_{11}i_3 + Z_{12}i_4$$
$$v_4 = Z_{21}i_3 + Z_{22}i_4$$

The analysis of the diagram in Figure 3.8 shows that:

$$v_1 = v_4 \text{ and } v_2 = v_3$$

$$i_1 = -i_4 \text{ and } i_2 = -i_3$$

When currents i_3 and i_4 are replaced by their equivalent expressions in i_1 and i_2, and voltages v_3 and v_4 by their equivalent expressions in v_2 and v_3, the relation in impedance for the passive quadripole can be written as:

$$v_2 = -Z_{11}i_2 - Z_{12}i_1$$
$$v_1 = -Z_{21}i_2 - Z_{22}i_1$$

The expressions of currents i_1 and i_2 are given by the definition of the active quadripole using parameters Y_{ij}. When these two currents (i_1 and i_2) are replaced by their equivalent expression in the above relation, this yields:

$$v_2 = -Z_{11}(Y_{21}v_1 + Y_{22}v_2) - Z_{12}(Y_{11}v_1 + Y_{12}v_2)$$
$$v_1 = -Z_{21}(Y_{21}v_1 + Y_{22}v_2) - Z_{22}(Y_{11}v_1 + Y_{12}v_2)$$

Hence:

$$v_1 \left[Z_{11}.Y_{21} + Z_{12}.Y_{11} \right] + v_2 \left[1 + Z_{11}.Y_{22} + Z_{12}Y_{12} \right] = 0$$
$$v_1 \left[1 + Z_{21}.Y_{21} + Z_{22}Y_{11} \right] + v_2 \left[Z_{21}.Y_{22} + Z_{22}Y_{12} \right] = 0$$

It is a homogeneous system of equations that admits nonzero solutions v_1 and v_2, provided that the determinant (Δ) of this system is zero.

It is worth noting that when the passage of just one frequency through the feedback circuit is given preference, v_1 and v_2 are sinusoidal signals.

The active element is a nonlinear device and may present parasitic elements. The circuit is designed to generate a sine wave of angular frequency ω_0 (frequency $f_0 = \omega_0/2\pi$). This requires the use of complex notation.

– Calculation of determinant (Δ)

$$\Delta = \begin{vmatrix} Z_{11}.Y_{21} + Z_{12}.Y_{11} & 1 + Z_{11}.Y_{22} + Z_{12}Y_{12} \\ 1 + Z_{21}.Y_{21} + Z_{22}Y_{11} & Z_{21}.Y_{22} + Z_{22}Y_{12} \end{vmatrix} = 0$$

$$\Delta = \left[(Z_{11}.Y_{21} + Z_{12}.Y_{11})(Z_{21}.Y_{22} + Z_{22}Y_{12}) \right] -$$
$$\left[(1 + Z_{21}.Y_{21} + Z_{22}Y_{11})(1 + Z_{11}.Y_{22} + Z_{12}Y_{12}) \right] = 0$$

$$\Delta = \underline{Z_{11}.Z_{21}Y_{21}.Y_{22}} + Z_{11}.Z_{22}.Y_{12}Y_{21} + Z_{12}.Z_{21}Y_{11}.Y_{22} + \underline{Z_{12}.Z_{22}.Y_{11}.Y_{12}}$$
$$-1 - Z_{21}.Y_{21} - Z_{22}Y_{11} - Z_{11}.Y_{22} - Z_{12}Y_{12} - \underline{Z_{11}.Z_{21}.Y_{21}.Y_{22}} - Z_{12}.Z_{21}.Y_{21}.Y_{12}$$
$$-Z_{11}.Z_{22}Y_{11}.Y_{22} - \underline{Z_{12}.Z_{22}Y_{11}.Y_{12}} = 0$$

The underlined terms cancel each other out. Therefore, the determinant "Δ" has the following expression:

$$\Delta = Z_{11}.Z_{22}.Y_{12}Y_{21} + Z_{12}.Z_{21}Y_{11}.Y_{22} - 1 - Z_{21}.Y_{21} - Z_{22}Y_{11}$$
$$-Z_{11}.Y_{22} - Z_{12}Y_{12} - Z_{12}.Z_{21}.Y_{21}.Y_{12} - Z_{11}.Z_{22}Y_{11}.Y_{22} - = 0$$

knowing that the matrix [Z] of a quadripole is given by:

$$[Z] = \begin{bmatrix} Z_{11} & Z_{12} \\ Z_{21} & Z_{22} \end{bmatrix} \Rightarrow \Delta Z = Z_{11}.Z_{22} - Z_{12}.Z_{21}$$

The same applies to parameters Y_{ij}. The matrix [Y] and the determinant are defined as follows:

$$[Y] = \begin{bmatrix} Y_{11} & Y_{12} \\ Y_{21} & Y_{22} \end{bmatrix} \Rightarrow \Delta Y = Y_{11}.Y_{22} - Y_{12}.Y_{21}$$

The product $\Delta Z.\Delta Y$ is defined by the following expression:

$$\mathbf{\Delta Z.\Delta Y} = (Z_{11}.Z_{22} - Z_{12}.Z_{21})(Y_{11}.Y_{22} - Y_{12}.Y_{21})$$
$$\mathbf{\Delta Z.\Delta Y} = Z_{11}.Z_{22}.Y_{11}.Y_{22} + Z_{12}.Z_{21}Y_{12}.Y_{21} - Z_{11}.Z_{22}.Y_{12}.Y_{21} - Z_{12}.Z_{21}.Y_{11}.Y_{22}$$

Resuming the definition equation of determinant "Δ", the presence of the product represented by $\mathbf{\Delta Z.\Delta Y}$ can be noted.

The elements that define this product $\mathbf{\Delta Z.\Delta Y}$ are underlined below.

$$\Delta = \underline{Z_{11}.Z_{22}.Y_{12}Y_{21}} + \underline{Z_{12}.Z_{21}Y_{11}.Y_{22}} - 1 - Z_{21}.Y_{21} - Z_{22}Y_{11}$$
$$-Z_{11}.Y_{22} - Z_{12}Y_{12} - \underline{Z_{12}.Z_{21}.Y_{12}.Y_{21}} - \underline{Z_{11}.Z_{22}Y_{11}.Y_{22}} = 0$$

The operating equation of the oscillator is consequently obtained:

$$1 + Z_{21}.Y_{21} + Z_{22}Y_{11} + Z_{11}.Y_{22} + Z_{12}Y_{12} + \Delta Z.\Delta Y = 0$$

In order to find the oscillation frequency, the imaginary part (or at least the part that contains the angular frequency ω and not the element that contains the amplification parameter) should be cancelled out.

In order to find the condition for sustained oscillation, the real part of the operating equation of the oscillator should be cancelled out. The feedback quadripole is a passive linear quadripole, therefore: $Z_{21} = Z_{12}$.

3.3.2. *Example of passive linear quadripole*

To illustrate that $Z_{12} = Z_{21}$ for a passive linear quadripole, let us consider the example shown in Figure 3.9.

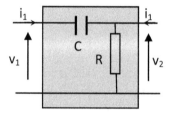

Figure 3.9. *Example of linear passive quadripole*

The definition of "R-C" quadripole by its impedance parameters is given by the following system of equations:

$$v_1 = Z_{11}i_1 + Z_{12}i_2$$
$$v_2 = Z_{21}i_2 + Z_{22}i_2$$

The capacitor impedance is equal to Z_C. Subsequently, it can be written:

$$v_1 = (Z_C + R)i_1 + Ri_2$$
$$v_2 = Ri_1 + Ri_2$$

As a result of identification, it can be found that:

$$Z_{11} = (Z_C + R); \qquad Z_{12} = Z_{21} = R \qquad \text{and} \qquad Z_{22} = R$$

This simple example can be used to verify that the passive linear quadripole indeed presents the characteristic: $Z_{12} = Z_{21}$.

Thus, for any passive linear quadripole, the following equalities can be stated:

$$Z_{12} = Z_{21} \quad \text{and} \quad Y_{12} = Y_{21}$$

3.4. Oscillator case study: Colpitts oscillator

3.4.1. *Presentation*

The feedback quadripole in case of a Colpitts oscillator (Figure 3.10) is a passive linear quadripole. The operating equation of the oscillator is defined by:

$$1 + Z_{12}.(Y_{12} + Y_{21}) + Z_{22}Y_{11} + Z_{11}.Y_{22} + \Delta Z.\Delta Y = 0$$

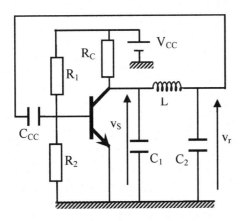

Figure 3.10. *Colpitts oscillator*

Parameters $\mathbf{Z_{ij}}$ represent the passive quadripole and parameters $\mathbf{Y_{ij}}$ represent the active quadripole.

It is also possible to define the passive quadripole by means of its admittance parameters. This can be done using the relations of passage between impedance and admittance parameters and rewriting the operating equation of the oscillator.

Similar to the active quadripole, hybrid parameters could be employed instead of admittance parameters (this can be done when the transition frequency f_T is very high compared to oscillation frequency f_0). In this case, too, it is sufficient to use the relations of passage between these various parameters.

The transition frequency of the chosen transistor is assumed to be very high compared with the frequency of the signal to be generated. This hypothesis is simply needed in order to simplify the calculations and facilitate the comprehension and knowledge acquisition of how high-frequency oscillators operate. The equivalent diagram of the transistor is therefore limited to the use of hybrid parameters.

The transistor is connected in "common emitter" configuration. For the sake of simplicity, parameter h_{12} is considered zero (which is practically the case).

Resistances R_1, R_2 and R_C serve for the proper polarization of the bipolar transistor.

The capacitor C_{CC} is considered a short-circuit at working frequency. Its role is to ensure that the polarization of the collector does not disturb the polarization of the base or vice versa, given that in static mode self-inductance L is a short circuit.

3.4.2. Operating equation of Colpitts oscillator

The amplifier is defined by its hybrid parameters and the feedback circuit is defined by its impedance parameters.

The equivalent diagram of the oscillator circuit is shown in Figure 3.11.

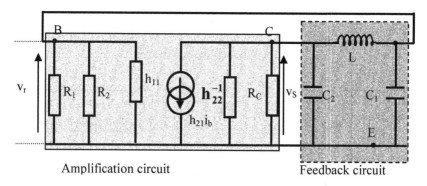

Amplification circuit   Feedback circuit

Figure 3.11. *Equivalent diagram of Colpitts oscillator*

The equivalent diagram of the oscillator circuit can be simplified as shown in Figure 3.12.

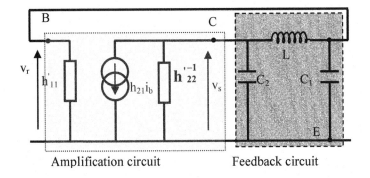

Amplification circuit   Feedback circuit

Figure 3.12. *Simplified diagram of Colpitts oscillator*

with:

$$h'_{11} = R_1 // R_2 // h_{11} = \frac{R_1 R_2 h_{11}}{R_1 R_2 + R_1 h_{11} + R_2 h_{11}}$$

$$h'_{22} == \frac{R_C h_{22} + 1}{R_C}$$

In the operating equation of the oscillator, the active quadripole is defined by its admittance parameters. The relations that ensure the passage between hybrid parameters and admittance parameters should be used in order to find the new operating equation of the oscillator.

The relations for passage from admittance parameters to hybrid parameters and conversely are as follows:

$$h_{11} = \left(\frac{v_1}{i_1}\right)_{v_2=0} = \frac{1}{Y_{11}} \qquad Y_{11} = \left(\frac{i_1}{v_1}\right)_{v_2=0} = \frac{1}{h_{11}}$$

$$h_{12} = \left(\frac{v_1}{v_2}\right)_{i_1=0} = -\frac{Y_{12}}{Y_{11}} \qquad Y_{12} = \left(\frac{i_1}{v_2}\right)_{v_1=0} = -\frac{h_{12}}{h_{11}}$$

$$h_{21} = \left(\frac{i_2}{i_1}\right)_{v_2=0} = \frac{Y_{21}}{Y_{11}} \qquad\qquad Y_{21} = \left(\frac{i_2}{v_1}\right)_{v_2=0} = \frac{h_{21}}{h_{11}}$$

$$h_{22} = \left(\frac{v_2}{i_2}\right)_{i_1=0} = \frac{\Delta Y}{Y_{11}} \quad \text{and} \quad Y_{22} = \left(\frac{i_2}{v_2}\right)_{v_1=0} = \frac{\Delta h}{h_{11}}$$

The operating equation then becomes:

$$1 + Z_{12}.(\frac{h_{21}}{h_{11}}) + Z_{22}\frac{1}{h_{11}} + Z_{11}.\frac{\Delta h}{h_{11}} + \Delta_Z \frac{h_{22}}{h_{11}} = 0$$

with: $\Delta h = h_{11}.h_{22} - h_{12}.h_{21} = h_{11}.h_{22}$

The operating equation of the oscillator can thus be deduced.

$$h_{11} + Z_{12}.h_{21} + Z_{22} + Z_{11}.\Delta h + \Delta_Z.h_{22} = 0$$

For the circuit that is of interest here, and in order to take into account the polarization resistances, h_{11} should be replaced by h'_{11} and h_{22} by h'_{22} with:

$$h'_{11} = \frac{R_1 R_2 h_{11}}{R_1 R_2 + R_1 h_{11} + R_2 h_{11}} \quad \text{and} \quad h'_{22} == \frac{R_C h_{22} + 1}{R_C}$$

These two parameters are real.

If the amplifier is defined by its hybrid parameters and the feedback circuit is defined by its admittance parameters, the relations of passage between admittance and impedance parameters of a quadripole, and conversely, should be used in order to redefine the operating equation of the oscillator.

This work aims at facilitating the calculation of parameters that are most of all related to the passive feedback quadripole if each time a bipolar transistor whose transition frequency is very high compared to the intended oscillation frequency is used.

The relations for the passage between impedance and admittance parameters are as follows:

$$Y_{11} = \left(\frac{Z_{22}}{\Delta Z}\right) \qquad\qquad Z_{11} = \left(\frac{Y_{22}}{\Delta Y}\right)$$

$$Y_{12} = -\left(\frac{Z_{12}}{\Delta Z}\right) \qquad\qquad Z_{12} = \left(\frac{-Y_{12}}{\Delta Y}\right)$$

$$Y_{21} = \left(\frac{-Z_{21}}{\Delta Z}\right) \qquad\qquad Z_{21} = -\left(\frac{Y_{21}}{\Delta Y}\right)$$

$$Y_{22} = \left(\frac{Z_{11}}{\Delta Z}\right) \qquad\qquad Z_{22} = \left(\frac{Y_{11}}{\Delta Y}\right)$$

When parameters Y_{ij} are used for the feedback quadripole and hybrid parameters h_{ij} are used for the amplification quadripole while taking into account polarization resistances, the operating equation of a Colpitts oscillator is defined by the following relation:

$$h'_{22} - Y_{12}.h_{21} + Y_{11} + Y_{22}.\Delta h' + \Delta_Y.h'_{11} = 0$$

3.4.3. Parameters of feedback quadripole

The feedback circuit can be schematically represented as shown in Figure 3.13.

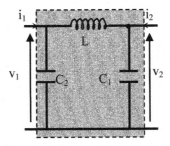

Figure 3.13. Feedback circuit in a Colpitts oscillator

The active quadripole is defined by its corrected parameters h_{ij} (covering polarization resistances).

The feedback quadripole can be defined by its impedance parameters (Z_{ij}) or by its admittance parameters in order to find the operating equation of the Colpitts oscillator and to define its operating characteristics, namely the oscillation frequency and the condition for sustained oscillation. Based on the diagram in Figure 3.13, the following can be written:

$$i_1 = Y_{11}v_1 + Y_{12}v_2$$
$$i_2 = Y_{21}v_1 + Y_{22}v_2$$

In sinusoidal mode:

$$i_1 = jC_2\omega v_1 + \frac{v_1 - v_2}{jL\omega} \qquad i_1 = j(C_2\omega - \frac{1}{jL\omega})v_1 - \frac{v_2}{jL\omega}$$

$$i_2 = jC_1\omega v_2 + \frac{v_2 - v_1}{jL\omega} \qquad i_2 = -\frac{v_1}{jL\omega} + j(C_1\omega - \frac{1}{L\omega})v_2$$

By identification, the following are obtained:

$$Y_{11} = jC_2\omega + \frac{1}{jL\omega} \quad \text{and} \quad Y_{12} = \frac{-1}{jL\omega}$$

$$Y_{21} = \frac{-1}{jL\omega} \quad \text{and} \quad Y_{22} = jC_1\omega + \frac{1}{jL\omega}$$

3.4.4. Oscillation frequency and condition for sustained oscillation

The operating equation of the oscillator is defined by:

$$h_{22}' - Y_{12}.h_{21} + Y_{11} + Y_{22}.\Delta h' + \Delta_Y.h_{11}' = 0$$

with: $\Delta Y = Y_{11}Y_{22} - Y_{12}Y_{21} = \dfrac{C_1\omega + C_2\omega - LC_1C_2\omega^3}{L\omega}$

The operating equation as a function of the parameters of the feedback circuit can be written as follows:

$$h_{22}' + \frac{1}{jL\omega}h_{21} + (jC_2\omega + \frac{1}{jL\omega}) + (jC_1\omega + \frac{1}{jL\omega}).\Delta h' +$$
$$\frac{C_1\omega + C_2\omega - LC_1C_2\omega^3}{L\omega}.h_{11}' = 0$$

$$jL\omega h_{22}' + h_{21} + (-LC_2\omega^2 + 1) + (-LC_1\omega^2 + 1).\Delta h' +$$
$$j(C_1\omega + C_2\omega - LC_1C_2\omega^3)h_{11}' = 0$$

3.4.4.1. Oscillation frequency

Cancelling out of the imaginary part yields the oscillation frequency, and cancelling out of the real part yields the condition for sustained oscillation.

$$f_0 = \frac{\omega_0}{2\pi} = \frac{1}{2\pi}\sqrt{\frac{Lh_{22}' + (C_1 + C_2)h_{11}'}{LC_1C_2h_{11}'}}$$

Admittance h'_{22} is generally very low and can be neglected ($h'_{22} \cong 0$). This allows the deduction of the oscillation frequency of the Colpitts oscillator as a function of the parameters of the feedback circuit.

$$f_0 = \frac{\omega_0}{2\pi} = \frac{1}{2\pi}\sqrt{\frac{(C_1 + C_2)}{LC_1C_2}} = \frac{1}{2\pi}\sqrt{\frac{1}{LC_{eq}}}$$

with: $C_{eq} = \dfrac{C_1.C_2}{(C_1 + C_2)}$

3.4.4.2. Condition for sustained oscillation

Canceling out of the real part yields:

$$h_{21} + (-LC_2\omega^2 + 1) + (-LC_1\omega^2 + 1).\Delta h' = 0$$

If $h_{12} = 0$ and $h'_{22} \cong 0$, $\Delta h' \cong 0$.

$$h_{21} + (-LC_2\omega_0^2 + 1) ; \text{ and } h_{21} + (-\frac{C_1 + C_2}{C_1} + 1) = 0$$

$$h_{21} - \frac{C_2}{C_1} = 0$$

Thus, the gain in transistor current should satisfy the following condition for sustained oscillation:

$$h_{21} = \frac{C_2}{C_1}$$

For the oscillator to effectively operate, the current gain is chosen as follows:

$$h_{21} > \frac{C_2}{C_1}$$

For the study of the Colpitts oscillator, the focus has been on determining the oscillation frequency and the condition for sustained oscillation.

These two parameters exist for steady state.

The study of transient state (start of oscillation and stabilization at given amplitude) requires nonlinear theory and introduces slightly more complex concepts that will be approached in chapter 4 of this book.

3.5. Hartley oscillator

3.5.1. *Schematic diagram*

The Hartley oscillator is schematically represented by the circuit in Figure 3.14. In order to simplify the study, the elements of polarization have been omitted.

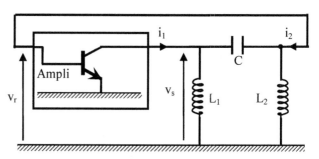

v_S: output voltage
v_r: voltage returned
from output to input
(positive feedback)

Figure 3.14. *Schematic diagram of the Hartley oscillator*

Similar to Colpitts oscillator, they can be readily integrated.

The amplifier involves a bipolar transistor connected to a common emitter whose transition frequency is very high compared with the oscillation frequency.

This amplifier element can then be represented by its hybrid parameters, while the feedback circuit (Figure 3.15) will be represented by its admittance parameters.

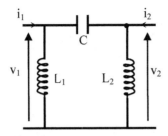

Figure 3.15. *Diagram of the feedback circuit of Hartley oscillator*

3.5.2. *Parameters of the feedback circuit*

It is worth recalling that the definition of parameters Y_{ij} is given by the following system of equations:

$$i_1 = Y_{11}v_1 + Y_{12}v_2$$
$$i_2 = Y_{21}v_1 + Y_{22}v_2$$

Since this is a sinusoidal oscillator, the following can be written:

$$i_1 = \frac{v_1}{jL_1\omega_1} + jC\omega(v_1 - v_2)$$

$$i_2 = jC\omega(v_2 - v_1) + \frac{v_2}{jL_2\omega}$$

The arrangement of these equations yields:

$$i_1 = v_1\left(jC\omega + \frac{1}{jL_1\omega}\right) - jC\omega v_2$$

$$i_2 = -jC\omega v_1 + v_2\left(jC\omega + \frac{1}{jL_2\omega}\right)$$

In order to find the parameters $\mathbf{Y_{ij}}$, it is sufficient to apply identification, which leads to:

$$Y_{11} = jC\omega + \frac{1}{jL_1\omega} \quad \text{and} \quad Y_{12} = -jC\omega$$

$$Y_{21} = -jC\omega \quad \text{and} \quad Y_{22} = jC\omega + \frac{1}{jL_2\omega}$$

3.5.3. *Operating equation*

Parameter h_{12} is considered equal to zero. Therefore, the following system of equations can be written for the amplifier element:

$$v_{r1} = -h_{11}i_2 + h_{12}v_S \qquad\qquad v_{r1} = -h_{11}i_2$$
$$-i_1 = h_{21}v_2 + h_{22}v_S \qquad\qquad -i_1 = h_{21}v_2 + h_{22}v_S$$

The operating equation of the oscillator is defined by:

$$h_{22} - Y_{12}.h_{21} + Y_{11} + Y_{22}.\Delta h + \Delta_Y.h_{11} = 0$$

$$\Delta Y = Y_{11}Y_{22} - Y_{12}Y_{21} = \frac{C}{L_1} + \frac{C}{L_2} - \frac{1}{L_1 L_2 \omega^2}$$

The operating equation as a function of the parameters of the passive feedback circuit is given by:

$$h_{22} + jC\omega h_{21} + (jC\omega + \frac{1}{jL_1\omega}) + (jC\omega + \frac{1}{jL_2\omega}).\Delta h +$$
$$(\frac{C}{L_1} + \frac{C}{L_2} - \frac{1}{L_1 L_2 \omega^2})h_{11} = 0$$

This yields:

$$jL_1 L_2 \omega h_{22} - L_1 L_2 C\omega^2 h_{21} + (-L_1 L_2 C\omega^2 + L_2) + (-L_1 L_2 C\omega^2 + L_1).\Delta h +$$
$$jL_1 L_2 C\omega h_{11}(\frac{C}{L_1} + \frac{C}{L_2} - \frac{1}{L_1 L_2 \omega^2}) = 0$$

3.5.4. *Oscillation frequency*

In order to find the oscillation frequency, it is sufficient to cancel out the imaginary part of the operating equation of the oscillator:

$$L_1 L_2 \omega_0 h_{22} + L_1 L_2 C\omega_0 h_{11}(\frac{C}{L_1} + \frac{C}{L_2} - \frac{1}{L_1 L_2 \omega_0^2}) = 0$$

Consequently, the oscillation frequency $\mathbf{f_0}$ can be defined as follows:

$$f_0 = \frac{\omega_0}{2\pi} = \frac{1}{2\pi}\sqrt{\frac{1}{[C(L_1 + L_2)] - \dfrac{h_{22}L_1L_2}{h_{11}C}}}$$

If parameter h_{22} is zero, then:

$$f_0 = \frac{\omega_0}{2\pi} = \frac{1}{2\pi}\sqrt{\frac{1}{[C(L_1 + L_2)]}}$$

3.5.5. *Condition for sustained oscillation*

The condition for sustained oscillation can be found by cancelling out the real part of the operating equation of the oscillator.

$$-L_1L_2C\omega_0^2h_{21} + (-L_1L_2C\omega_0^2 + L_2) + (-L_1L_2C\omega_0^2 + L_1).\Delta h = 0$$

$$h_{21} = (-1 + \frac{1}{L_1C\omega_0^2}) + (-1 + \frac{1}{L_2C\omega_0^2})\Delta h$$

$$\omega_0 = \sqrt{\frac{1}{[C(L_1 + L_2)] - \dfrac{h_{22}L_1L_2}{h_{11}C}}}$$

If h_{22} is assumed small and can be neglected (which is valid in most cases), then:

$$\Delta h = 0 \quad \text{and} \quad \omega_0 = \sqrt{\frac{1}{[C(L_1 + L_2)]}}$$

$$h_{21} = (-1 + \frac{1}{L_1C\omega_0^2})$$

This leads to the condition that the current gain of the bipolar transistor has to satisfy in order to have sustained oscillations.

$$h_{21} = \frac{L_2}{L_1}$$

3.6. Clapp oscillator

3.6.1. *Schematic diagram*

The schematic diagram of a Clapp oscillator circuit is shown in Figure 3.16. The polarization resistances of the bipolar transistor that serves as an amplifier element have been omitted.

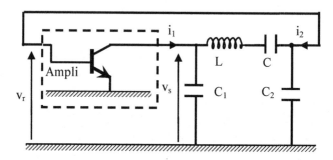

Figure 3.16. *Schematic diagram of Clapp oscillator*

The approach is similar to the one taken for the two previously studied oscillators (Colpitts and Hartley oscillators).

The active element is represented by its hybrid parameters and the passive element (feedback circuit) is represented by its admittance parameters.

For the active element, parameters h_{12} and h_{22} are considered negligible ($h_{12} \cong 0$ and $h_{22} \cong 0$).

3.6.2. *Operating equation*

The operating equation of the oscillator is defined by:

$$-Y_{12}.h_{21} + Y_{11} + \Delta Y.h_{11} = 0$$

Parameters h_{21} (gain in current) and h_{11} (internal impedance of the transistor) are given by the manufacturer.

It remains to determine the parameters Y_{ij} of the passive feedback circuit schematically represented in Figure 3.17, in order to fully define the operating equation.

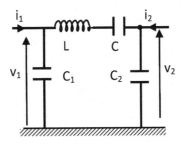

Figure 3.17. *Feedback circuit of Clapp oscillator*

In sinusoidal mode, the following can be written:

$$i_1 = jC_1\omega v_1 + \frac{jC\omega}{1-LC\omega^2}(v_1 - v_2)$$

$$i_2 = jC_2\omega v_1 + \frac{jC\omega}{1-LC\omega^2}(v_2 - v_1)$$

The arrangement of this system of equations yields:

$$i_1 = v_1.j(C_1\omega + \frac{C\omega}{1-LC\omega^2}) - v_2\frac{jC\omega}{1-LC\omega^2}$$

$$i_2 = -v_1\frac{jC\omega}{1-LC\omega^2} + v_2.j(C_2\omega + \frac{C\omega}{1-LC\omega^2})$$

By identification, the expressions of various parameters Y_{ij} can readily be obtained.

$$Y_{11} = j(C_1\omega + \frac{C\omega}{1-LC\omega^2}) \qquad Y_{12} = -\frac{jC\omega}{1-LC\omega^2}$$

$$Y_{22} = j(C_2\omega + \frac{C\omega}{1-LC\omega^2}) \qquad Y_{21} = -\frac{jC\omega}{1-LC\omega^2}$$

It should be recalled that the operating equation of the oscillator according to the formulated hypotheses satisfies:

$$-Y_{12}.h_{21} + Y_{11} + \Delta Y.h_{11} = 0 \text{ with } \Delta Y = Y_{11}Y_{22} - Y_{12}Y_{21}$$

$$\Delta Y = -(C_1\omega + \frac{C\omega}{1-LC\omega^2})(C_2\omega + \frac{C\omega}{1-LC\omega^2}) + \frac{(C\omega)^2}{(1-LC\omega^2)^2}$$

$$\Delta Y = -(C_1C_2\omega^2 + \frac{C_1C\omega^2}{1-LC\omega^2} + \frac{C_2C\omega^2}{1-LC\omega^2})$$

The operating equation as a function of the parameters of the feedback circuit is then:

$$\frac{C\omega}{1-LC\omega^2}.h_{21} + (C_1\omega + \frac{C\omega}{1-LC\omega^2}) +$$

$$j(C_1C_2\omega^2 + \frac{C_1C\omega^2}{1-LC\omega^2} + \frac{C_2C\omega^2}{1-LC\omega^2}).h_{11} = 0$$

3.6.3. *Characteristic parameters of the oscillator*

3.6.3.1. *Oscillation frequency*

Oscillation frequency f_0 (or angular frequency ω_0) is obtained by cancelling out the imaginary part.

$$(C_1C_2\omega_0^2 + \frac{C_1C\omega_0^2}{1-LC\omega_0^2} + \frac{C_2C\omega_0^2}{1-LC\omega_0^2}) = 0$$

$$\left[(1-LC\omega_0^2)C_1C_2 + C_1C + C_2C\right]\omega_0^2 = 0$$

Solution $\omega_0 = 0$ should be rejected. The following can then be written:

$$\left[(1-LC\omega_0^2)C_1C_2 + C_1C + C_2C\right] = 0$$

Based on this equation, the angular frequency ω_0 or the oscillation frequency f_0 can be deduced.

$$\omega_0 = \sqrt{\frac{C_1C_2 + C_1C + C_2C}{LC_1C_2}}; \quad f_0 = \frac{1}{2\pi}\sqrt{\frac{1}{L}\left[\frac{1}{C_1} + \frac{1}{C_2} + \frac{1}{C}\right]}$$

3.6.3.2. *Condition for sustained oscillation*

In order to find the condition for sustained oscillation, it is sufficient to cancel out the real part of the operating equation at the oscillation frequency.

$$\frac{C\omega_0}{1-LC\omega_0^{\,2}}.h_{21} + (C_1\omega_0 + \frac{C\omega_0}{1-LC\omega_0^{\,2}}) = 0$$

This leads to:

$$h_{21} = \frac{C_1}{C_2}$$

It should be recalled that this value of the gain in current of the bipolar transistor is a minimum value for the oscillation to start.

3.7. Quartz crystal oscillator

Quartz crystal is used in oscillators when there is a need to generate a wave with very high-frequency stability.

All the devices that require very stable frequencies in order to properly operate use quartz crystal oscillators (for example, microcomputers and emission and reception devices in telecommunication). Quartz crystal stability is due to its very high-quality "Q" factor ($Q \approx 10^6$).

3.7.1. *Frequency stability of an oscillator*

In order to generate a sustained sinusoidal wave, an oscillator must necessarily satisfy the Barkhausen condition and the structure schematically shown in Figure 3.18.

The amplification part is represented by the voltage gain "A" and the feedback circuit is represented by its transfer function "$\beta(\omega)$". For the sake of simplicity, the gain "A" is supposed to be positive real.

Barkhausen condition is defined by: A.$\beta(\omega)$= 1. As already explained, this relation contains a condition for the module and a condition for the phase.

$$A.\beta = 1 \Leftrightarrow \begin{array}{c} Arg(A)+Arg(\beta) = 0 \quad [2\pi] \\ |A.\beta| = 1 \end{array}$$

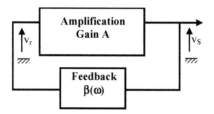

Figure 3.18. *Structure of feedback oscillator*

Assuming that the gain is positive real, the focus will be on the study of the argument "φ" of the feedback circuit β(ω).

This phase can vary due to several causes:

– poor insulation of the oscillator circuit, parasitic capacitances and influence of temperature;

– aging of components and variation in their characteristics.

The assumption is made that there are two possibilities for the variations in the argument "φ" of the feedback circuit.

First case: Slow variation in "φ" around the oscillation frequency f_0 (Figure 3.19). The slope (dφ/df) is small around f_0.

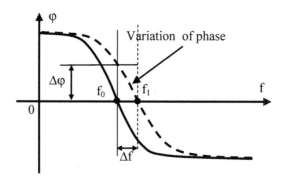

Figure 3.19. *Variation of the argument of the transfer function of the feedback circuit of a harmonic oscillator around the oscillation frequency*

When for whatever reason there is a variation in the argument of the transfer function of the feedback circuit, there is an incidental variation in the oscillation frequency, which ensures that the Barkhausen condition is still verified, and the oscillator delivers a sinusoidal wave. It can be seen that when the slope ($d\varphi/df$) is small around f_0, the oscillation frequency goes from f_0 to f_1. This variation $\Delta f = f_1 - f_0$ can be very harmful. Δf should be reduced as much as possible, so that the oscillation frequency is maintained as close as possible to f_0 and thus ensure its stability.

Second case: "φ" varies rapidly around the oscillation frequency (Figure 3.20). The slope ($d\varphi/df$) is large around f_0.

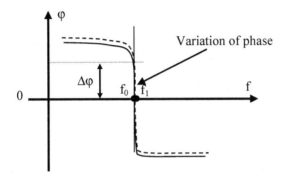

Figure 3.20. *Rapid variation of the argument of the transfer function of the feedback circuit around f_0.*

As can be noted in this case, when the phase related to the argument of the transfer function of the feedback circuit varies rapidly around the oscillation frequency, the stability of this oscillation frequency is reached. Frequencies f_0 and f_1 are practically identical.

$$\Delta f = f_1 - f_0 \rightarrow 0$$

Thus, the frequency stability of an oscillator can be defined by the expression:

$$S = f_0 \frac{d\phi(f)}{df}$$

The size of S is an indicator of the stability of the oscillation frequency "f_0" of an oscillator. An oscillator with good frequency stability should necessarily have a feedback circuit with a phase that varies very rapidly around the oscillation frequency. The larger the slope of the phase around "f_0", the better the frequency stability of the oscillator.

3.7.2. *Quartz crystal operation*

3.7.2.1. *Piezoelectric effect*

Quartz crystal has a very high quality factor. It has the characteristic that when mechanical stress (compression or stretching) is applied to its two faces (Figure 3.21), a potential difference V is generated across it.

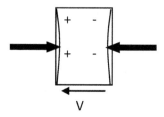

Figure 3.21. *When compression force is applied on the faces of a quartz crystal, a potential difference is generated*

On the other hand, when an electric potential difference is applied across the quartz crystal, the latter undergoes deformation. As soon as there is no voltage across it, the quartz crystal regains its initial shape.

These properties are known as the **piezoelectric effect.**

Owing to the piezoelectric effect, the quartz crystal behaves as a resonant "RLC" circuit.

3.7.2.2. *Equivalent diagram of a quartz crystal*

The equivalent electrical diagram of a quartz crystal is shown in Figure 3.22.

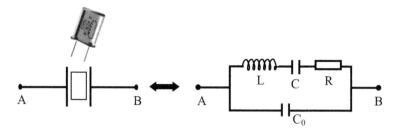

Figure 3.22. *Equivalent electrical diagram of quartz crystal*

The equivalent diagram of quartz crystal involves a series RLC circuit due to the piezoelectric effect and a parasitic parallel capacitance C_0 related to the connecting wire. Resistance R is related to mechanical losses, with the value of R approaching one hundred Ohms. In most cases, this resistance is neglected.

The value of capacitance C_0 is of several pico-Farad (10^{-12} Farad). The value of capacitance C is of the order of femto-Farad (10^{-15} Farad).

In general:

$$R << \frac{1}{2\pi F C_0}$$ F being the working frequency of quartz crystal.

3.7.3. Equivalent impedance of quartz crystal

The impedance of quartz crystal consists of two parallel impedances Z_1 and Z_2, as shown in Figure 3.23.

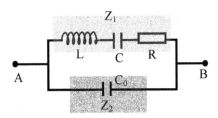

Figure 3.23. *Impedance of quartz crystal*

Thus, it can be written:

$$Z_{AB} = Z_1 // Z_2$$

with:

$$Z_1 = (R + jL\omega + \frac{1}{jC\omega}) \quad \text{and} \quad Z_2 = \frac{1}{jC_0\omega}$$

This leads to the expression of overall impedance of quartz crystal:

$$Z_{AB} = \frac{1}{\omega}\left[\frac{LC\omega^2 - 1 - jRC\omega}{RCC_0\omega + j(LCC_0\omega^2 - (C + C_0))}\right]$$

When the influence of resistance R is neglected, the expression of the impedance becomes:

$$Z_{AB} = -j\frac{1}{\omega}\left[\frac{LC\omega^2 - 1}{(LCC_0\omega^2 - (C + C_0))}\right]$$

It is a pure reactance. This impedance could also be written as follows:

$$Z_{AB} = -j\frac{1}{\omega}\left[\frac{1 - LC\omega^2}{(C + C_0)(1 - \frac{LCC_0\omega^2}{C + C_0})}\right];$$

$$Z_{AB} = -j\frac{1}{\omega}\left[\frac{1 - \frac{\omega^2}{\omega_S^2}}{(C + C_0)(1 - \frac{\omega^2}{\omega_P^2})}\right]$$

with:

$$\omega_S = \frac{f_S}{2\pi} = \sqrt{\frac{1}{LC}} \quad \text{and} \quad \omega_P = \frac{f_P}{2\pi} = \sqrt{\frac{C_0 + C}{LCC_0}}$$

The two frequencies that have been defined represent:

f_S – Series frequency: the resonance frequency of the series circuit formed by the two dipole devices related to self-inductance "L" and capacitor "C".

f_P – Parallel frequency: the resonance frequency of the parallel circuit formed by the dipole devices related to self-inductance "L" and capacitor "C" and capacitor "C_0".

The relation between the series frequency "f_S" and the parallel frequency "f_P" can be defined as follows:

$$f_S = \frac{1}{2\pi}\sqrt{\frac{1}{LC}} \quad \text{and} \quad f_P = \frac{1}{2\pi}\sqrt{\frac{C_0 + C}{LCC_0}}$$

$$\frac{f_S}{f_P} = \frac{\left[\dfrac{C_0 C}{LC + C_0}\right]^{\frac{1}{2}}}{(LC)^{\frac{1}{2}}} = \left[\frac{C_0}{C + C_0}\right]^{\frac{1}{2}}$$

The value of capacitance C is known to be very small compared with the value of capacitance C_0.

$$\frac{f_P}{f_S} = \left[\frac{C + C_0}{C_0}\right]^{\frac{1}{2}} = \left[1 + \frac{C}{C_0}\right]^{\frac{1}{2}} \cong 1 + \frac{C}{2C_0}$$

$$f_P \cong f_S\left[1 + \frac{C}{2C_0}\right]$$

Thus, it can be very easily noted that parallel frequency f_P is very close to series frequency f_S. It should be recalled that capacitance C is of the order of femto-Farad and that capacitance C_0 is of the order of pico-Farad. Then it can be deduced that:

$$f_P \cong f_S$$

Parallel frequency f_P is very close to series frequency f_s. However, frequency "f_P" is slightly above frequency "f_S".

3.7.4. Frequency behavior of a quartz crystal

As already mentioned, when the influence of resistance R is neglected, the impedance of a quartz crystal can be written as follows:

$$Z_{AB} = -j\frac{1}{\omega}\left[\frac{1 - \dfrac{\omega^2}{\omega_S^2}}{(C + C_0)(1 - \dfrac{\omega^2}{\omega_P^2})}\right] = -jX$$

The expression of reactance X is given by:

$$X = \frac{1}{\omega}\left[\frac{1 - \dfrac{\omega^2}{\omega_S^2}}{(C + C_0)(1 - \dfrac{\omega^2}{\omega_P^2})}\right]$$

The plot of the evolution of reactance "X" as a function of frequency is shown in Figure 3.24.

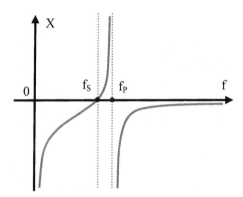

Figure 3.24. *Evolution of reactance "X" as a function of frequency*

It would also be interesting to plot the impedance of quartz crystal in terms of module and of phase (Figure 3.25). This graphical representation provides very interesting information on the benefits of using quartz crystal for stabilizing the frequency of the signal to be generated.

Indeed, much can be learned from the analysis of the variations in phase within the very narrow range of frequency between f_S and f_P.

It can be noted that in the frequency range $\Delta f = (f_P - f_S)$, the phase varies very rapidly. Δf is practically zero ($f_P \cong f_S$).

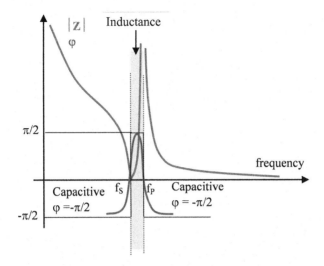

Figure 3.25. *Evolution of quartz crystal impedance in terms of module and phase. For a color version of this figure, see www.iste.co.uk/haraoubia/nonlinear1.zip*

Consequently, the quartz crystal has very high stability, since this is a situation where:

$$\frac{d\phi(f)}{df} \to \infty$$

This rapid variation in the phase of quartz crystal impedance between parallel frequency and series frequency is an incentive to use the quartz crystal in the vicinity of these frequencies.

The quartz crystal has various behaviors depending on the frequency range in which it is used. The analysis of quartz crystal impedance shows that it is possible to have three remarkable frequency ranges (Figure 3.26):

– $f < f_S$: quartz crystal is equivalent to a capacitance;

– $f_S < f < f_P$: quartz crystal is equivalent to a self-inductance;

– $f > f_P$: quartz crystal is equivalent to a capacitance.

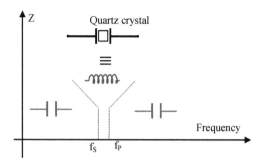

Figure 3.26. *Behavior of the quartz crystal depending on frequency evolution*

The series frequency "f_S" and the parallel frequency "f_P" are very close, if not virtually equal. Consequently, the frequency range within which the quartz crystal behaves as a self-inductance is very narrow. The self-inductance in question has a very high quality factor.

For the quartz crystal, since there is a series resonance and a parallel resonance; the series and parallel quality factors can be defined as follows:

$$Q_S = \frac{L\omega_S}{R} \quad \text{and} \quad Q_P = \frac{L\omega_P}{R}$$

As it is known that $\omega_s \cong \omega_p$, the following can be written:

$$\begin{aligned} Q_S &= \frac{L\omega_S}{R} \\ Q_P &= \frac{L\omega_P}{R} \end{aligned} \Rightarrow Q_S \cong Q_P$$

Example: Let us have a quartz crystal with the following characteristics:

L = 0.5H; C = 50fF; C_0 = 8pF

The calculation of the series frequency and of the parallel frequency yields:

f_S =1.007 MHz; f_P = 1.01 MHz; R several tens of Ohms.

It is possible to obtain a very high-quality "Q" factor for this type of quartz crystal, compared with a classic RLC circuit.

Therefore, due to its spectral purity properties, the quartz crystal can be used as a resonant circuit, for example, in a Colpitts oscillator.

3.7.5. *Example of quartz crystal oscillator*

Let us consider the circuit that is schematically shown in Figure 3.27. The amplifier element is a bipolar transistor. A quartz crystal has been inserted in the feedback circuit and the task at hand is the study of this device as an oscillator.

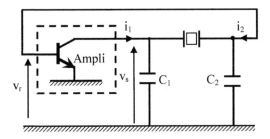

Figure 3.27. *Example of quartz crystal oscillator*

To facilitate comprehension and for reasons of simplicity, the transistor polarization elements have been omitted.

The objective is to determine the oscillation frequency and the condition for sustained oscillations of this oscillator.

It is possible to assume that transition frequency is very high compared with working frequency and to focus on using hybrid parameters for the bipolar transistor.

3.7.5.1. *Expressions of admittance parameters of the feedback circuit*

The diagram of the feedback circuit is shown in Figure 3.28. In the diagram of the quartz crystal, the effect of resistance R has been neglected.

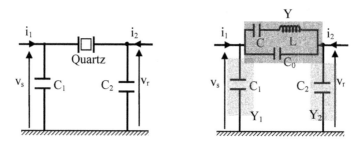

Figure 3.28. *Equivalent diagram of the feedback circuit in which a quartz crystal has been inserted*

Based on this diagram, the admittance parameters of the quadripole will be calculated.

$$i_1 = Y_1 v_S + Y(v_S - v_r)$$
$$i_2 = Y_2 v_r - Y(v_S - v_r)$$

with:

$$Y_1 = jC_1\omega; \qquad Y_2 = jC_2\omega;$$

$$Y = jC_0\omega + \frac{1}{Z}; \qquad Z = jL\omega + \frac{1}{jC\omega};$$

$$i_1 = (Y_1 + Y)v_S - Yv_r$$
$$i_2 = -Yv_r + (Y_2 + Y)v_S$$

The above-mentioned system of equations leads to the parameters Y_{ij} of the feedback quadripole:

$$Y_{11} = (Y_1 + Y); \qquad Y_{12} = -Y$$
$$Y_{21} = -Y; \qquad Y_{22} = (Y_2 + Y)$$

The expressions of admittance parameters as a function of the elements of the feedback circuit are as follows:

$$Y_{11} = j\omega(C_1 + C_0) + \frac{1}{Z}; \qquad Y_{12} = -(jC_0\omega + \frac{1}{Z});$$

$$Y_{21} = -(jC_0\omega + \frac{1}{Z}); \qquad Y_{22} = j\omega(C_2 + C_0) + \frac{1}{Z}$$

$$\Delta Y = Y_1 Y_2 + Y_1 Y + Y_2 Y = Y_1 Y_2 + Y(Y_1 + Y_2)$$

$$\Delta Y = \frac{j\omega}{Z}(C_1 + C_2) - \omega^2 \left[C_0(C_1 + C_2) + C_1 C_2 \right]$$

3.7.5.2. Operating equation and parameters of the quartz crystal

The operating equation of the oscillator is given by the following relation:

$$h_{22} - Y_{12}.h_{21} + Y_{11} + Y_{22}.\Delta h + \Delta Y.h_{11} = 0$$

If parameters h_{12} and h_{22} are assumed to be equal to zero, the operating equation of the oscillator is simplified and becomes:

$$-Y_{12}.h_{21} + Y_{11} + \Delta_Y.h_{11} = 0$$

$$Y_{11} = j\omega(C_1 + C_0) + \frac{1}{Z};$$

$$Y_{12} = Y_{21} = -(jC_0\omega + \frac{1}{Z});$$

$$\frac{1}{Z} = \frac{jC\omega}{1 - LC\omega^2}$$

Replacement of various parameters by their equivalent expression yields:

$$(jC_0\omega + \frac{1}{Z})h_{21} + j\omega(C_1 + C_0) + \frac{1}{Z} +$$

$$\left[\frac{j\omega}{Z}(C_1 + C_2) - \omega^2 \left[C_0(C_1 + C_2) + C_1 C_2 \right] \right] h_{11} = 0$$

$$-(jC_0\omega+\frac{jC\omega}{1-LC\omega^2})h_{21}+j\omega(C_1+C_0)+\frac{jC\omega}{1-LC\omega^2}$$

$$+\left[\frac{-C\omega^2}{1-LC\omega^2}(C_1+C_2)-\omega^2\left[C_0(C_1+C_2)+C_1C_2\right]\right]h_{11}=0$$

$$\omega(C_0+\frac{C}{1-LC\omega^2})h_{21}+\omega\left[(C_1+C_0)+\frac{C}{1-LC\omega^2}\right]$$

$$+\omega^2\left[\frac{-jC}{1-LC\omega^2}(C_1+C_2)-j\left[C_0(C_1+C_2)+C_1C_2\right]\right]h_{11}=0$$

In order to find the oscillation frequency, it is sufficient to cancel out the imaginary part of the operating equation of the oscillator.

$$\omega^2\left[\frac{C}{1-LC\omega^2}(C_1+C_2)+\left[C_0(C_1+C_2)+C_1C_2\right]\right]h_{11}=0$$

Solution $\omega=0$ should be rejected.

$$\left[\frac{C}{1-LC\omega_0^2}(C_1+C_2)+\frac{(1-LC\omega_0^2)\left[C_0(C_1+C_2)+C_1C_2\right]}{1-LC\omega_0^2}\right]h_{11}=0$$

Hence:

$$LC\omega_0^2=\frac{C_0C_1+C_0C_2+C_1C_2+CC_1+CC_2}{C_0C_1+C_0C_2+C_1C_2}$$

Finally, this yields the relation that defines the oscillation frequency:

$$\omega_0^2=\frac{1}{LC}+\frac{C_1+C_2}{L(C_0C_1+C_0C_2+C_1C_2)}$$

It should be recalled that C_0 is a parasitic capacitance. The following approximation can be written:

$$C_0(C_1+C_2)<<C_1C_2$$

Under these conditions, the relation that defines the oscillation frequency is given by:

$$\omega_0^2 = \frac{1}{LC} + \frac{C_1 + C_2}{LC_1C_2};$$

$$\omega_0^2 = \frac{1}{LC} + \frac{1}{LC_1} + \frac{1}{LC_2} = \frac{1}{L}\left[\frac{1}{C} + \frac{1}{C_1} + \frac{1}{C_2}\right]$$

Hence, it can be inferred that the feedback circuit that contains the quartz crystal comes down to a parallel resonant circuit formed of the self-inductance "L" of the quartz crystal and three capacitors C, C_1 and C_2 connected in series (Figure 3.29).

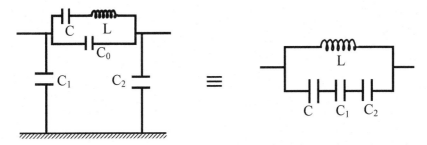

Figure 3.29. *Equivalence between the feedback circuit connected to a quartz crystal oscillator and the parallel resonant circuit*

It is worth recalling that the unit of measurement of capacitance C is the femto-farad (10^{-15} Farad). Therefore, this capacitance is very small compared with capacitances C_1 and C_2.

Hence, the oscillation frequency of the quartz crystal oscillator is defined as follows:

$$\omega_0 = \sqrt{\frac{1}{L}\left[\frac{1}{C} + \frac{1}{C_1} + \frac{1}{C_2}\right]} \cong \sqrt{\frac{1}{LC}};$$

$$f_0 = \frac{1}{2\pi}\sqrt{\frac{1}{LC}}$$

This frequency is none other than the series resonance frequency of quartz crystal.

In order to determine the condition for sustained oscillation, the real part of the operating equation of the oscillator should be cancelled out.

$$(C_0 + \frac{C}{1-LC\omega_0^2})h_{21} + \left[(C_1+C_0) + \frac{C}{1-LC\omega_0^2} \right] = 0$$

It should be recalled that:

$$\omega_0^2 = \frac{1}{L}\left[\frac{1}{C} + \frac{1}{C_1} + \frac{1}{C_2} \right]$$

Replacing ω_0 with the equivalent expression leads to:

$$(C_0 + \frac{C}{1-C.\left[\frac{1}{C}+\frac{1}{C_1}+\frac{1}{C_2}\right]})h_{21} + (C_1+C_0) + \frac{C}{1-C.\left[\frac{1}{C}+\frac{1}{C_1}+\frac{1}{C_2}\right]} = 0$$

$$(C_0 - \frac{C_1 C_2}{C_1+C_2})h_{21} + (C_1+C_0) - \frac{C_1 C_2}{C_1+C_2} = 0$$

$$(\frac{C_0(C_1+C_2)-(C_1 C_2)}{C_1+C_2})h_{21} + \frac{C_1^2 + C_0(C_1+C_2)}{C_1+C_2} = 0$$

Simplifications are possible, knowing that:

$$C_0(C_1+C_2) \ll C_1 C_2 \quad \text{and} \quad C_0(C_1+C_2) \ll C_1^2$$

Then:

$$(\frac{-(C_1 C_2)}{C_1+C_2})h_{21} + \frac{C_1^2}{C_1+C_2} = 0$$

Finally, the condition that current gain should meet in order to ensure the proper operation of the quartz crystal oscillator is:

$$h_{21} = \frac{C_1}{C_2}$$

This is the same condition as the one obtained for a Colpitts oscillator that uses an LC circuit in π as a feedback circuit.

The same statement is applicable to the current gain. The calculated gain is a minimum value.

The condition required for oscillations to start is:

$$h_{21} > \frac{C_1}{C_2}$$

Oscillator as a Nonlinear Device

4.1. Introduction

Let us recall that an oscillator is a loop device (Figure 4.1).

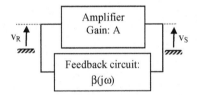

Figure 4.1. *Principle of feedback oscillator*

In order to have an oscillator, the condition for stable oscillations (Barkhausen condition) must be satisfied:

$$A\beta(j\omega) = 1$$

This condition cannot be verified very rigorously. Indeed, it is possible to have either:

$$A\beta(j\omega) = 1 - \varepsilon \text{ or } A\beta(j\omega) = 1 + \varepsilon$$

where ε is a positive real.

In the first case, the amplifier can be thought of as not being able to compensate the losses of the feedback circuit, and consequently, if an oscillation emerges, then its attenuation is an exponential decrease, as shown in Figure 4.2(a).

If $A\beta(j\omega) = 1 + \varepsilon$, then the oscillation increases until saturation (Figure 4.2(b)).

If ε is zero, then a sustained sinusoidal oscillation of angular frequency ω_0 is obtained (Figure 4.2(c)).

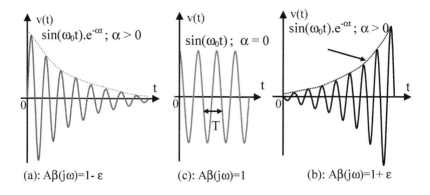

(a): $A\beta(j\omega)=1-\varepsilon$ (c): $A\beta(j\omega)=1$ (b): $A\beta(j\omega)=1+\varepsilon$

Figure 4.2. *Evolution of the output signal of an oscillator in relation with the Barkhausen condition. For a color version of this figure, see www.iste.co.uk/haraoubia/nonlinear1.zip*

Signal disappearance in case of a damped oscillation is not of significant interest for the study of the behavior and operation of oscillators.

However, when the signal is increasing ($A\beta > 1$), the amplifier output will go into saturation mode (signal peak clipping) and the amplifier will consequently leave its linear operation domain.

The output signal is no longer sinusoidal, but rather resembles a pseudo-rectangular signal (distortion of the output signal).

Given the above, the study of the oscillator should make use of nonlinear theory instead of linear theory.

An example of a method involving a nonlinear approach of oscillators is the first harmonic method.

When the Barkhausen condition is met, the power supplied by the amplifier compensates the feedback circuit losses, which can be represented as dissipation across resistor R (the reader is referred to the study on the RLC oscillator circuit in Chapter 3).

Therefore, the following three possibilities can be considered in terms of power:

– consumed power (P_C) is lower than supplied power (P_F):

$(P_C) < (P_F)$

– consumed power (P_C) is higher than supplied power (P_F):

$(P_C) > (P_F)$

– consumed power (P_C) is equal to supplied power (P_F):

$(P_C) = (P_F)$.

Since the focus is here on a sinusoidal oscillator, the output voltage has the following form: $v_S = V_M.\sin(2\pi f_0 t)$, where V_M is the peak voltage and f_0 is the oscillation frequency.

The power dissipated or consumed across the feedback circuit, which can be represented as loss resistance R, is given by the following relation:

$$P_C = \frac{1}{T}\int_0^T \frac{v_S^2(t)}{R}dt = \frac{1}{T}\int_0^T \frac{v_M^2.\sin^2(2\pi f_0 t)}{R}dt = \frac{1}{T}\int_0^T \frac{v_M^2.(1-\cos(4\pi f_0 t))}{2R}dt$$

$$P_C = \frac{v_M^2}{2R}$$

The representation of various cases of supplied power and consumed power is shown in Figure 4.3.

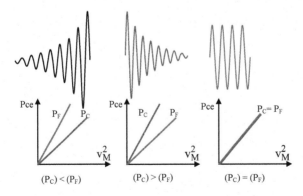

Figure 4.3. *Evolution of the signal at oscillator output as a function of powers supplied by the amplifier and consumed by the feedback circuit. For a color version of this figure, see www.iste.co.uk/haraoubia/nonlinear1.zip*

4.2. Stability of an oscillator

Let us recall that an oscillator consists of two quadripoles, an amplification circuit and a feedback circuit (passive linear circuit), which enables setting the oscillation frequency.

The power consumed (P_C) or dissipated across a loss resistance can be expected to vary linearly with the peak amplitude of the signal to be generated.

Given the mentioned reasons, the power supplied (P_F) by the amplifier cannot be linear.

Under these conditions, it is important to consider all the possibilities related to the (nonlinear) power supplied with respect to the (linear) power consumed.

The simplest case to be studied is when the supplied power is lower than the consumed power. In this case, oscillation cannot start.

Even if oscillation starts, it will very rapidly dampen, since it is not sustained.

4.2.1. *Static stability*

Static stability can be defined as energy balance between supplied power and consumed power in a given point.

Getting away from this point leads to full balance disruption.

In case of oscillators, at static equilibrium, oscillation is sustained.

As soon as the system gets away from the static equilibrium point, in either direction, the oscillation is no longer self-stabilized in terms of amplitude.

To further clarify this aspect, let us suppose that the supplied power (P_F) and consumed power (P_C) evolve according to Figure 4.4.

Supplied power is nonlinear, whereas consumed power is a linear function of the peak amplitude of the generated signal.

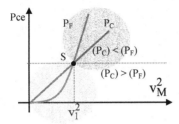

Figure 4.4. *S is the point of static equilibrium. For a color version of this figure, see www.iste.co.uk/haraoubia/nonlinear1.zip*

At point "S", supplied power is equal to consumed power. If the oscillator operation can be maintained at point "S", then the output will be a sustained sinusoidal oscillation. However, this cannot be permanent, since for various reasons, the components can drift and the equilibrium is disrupted in any direction (above or below point S). The following two areas should therefore be considered:

$V_M < V_1$: $P_F < P_C$

$V_M > V_1$: $P_F > P_C$

In both cases, the oscillation obtained has no constant stable amplitude. It is divergent (the amplifier goes to saturation) if $V_M > V_0$. The oscillation does not start if $V_M < V_1$.

This leads to the conclusion that point "S" is a point of energy balance and ensures only static stability.

Because of this, the oscillator based on the variation of supplied and consumed powers is rendered useless, since it does not meet the need to generate a sustained sinusoidal wave.

4.2.2. *Dynamic stability*

Dynamic stability can be defined as the energy balance between the supplied power and the consumed power at a given point. In contrast, with static stability, when getting away from this point, the balance is not fully disrupted. In fact, as shown by the analysis of the curves in Figure 4.5, at point D, oscillation is sustained. When getting away from point D, the following two directions are possible:

$V_M > V_1$: supplied power is lower than consumed power, and consequently, the output signal is damped until its peak amplitude reaches the value V_1.

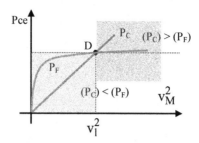

Figure 4.5. *Point D is a point of dynamic equilibrium. For a color version of this figure, see www.iste.co.uk/haraoubia/nonlinear1.zip*

$V_M < V_1$: supplied power is higher than consumed power, and consequently, the peak amplitude of the output signal increases and reaches the value V_1.

Thus, the amplitude of the generated signal is stabilized at the level of value V_1. Point D is a point of dynamic stability.

4.3. Nonlinear phenomena in oscillators

As already highlighted, the study of oscillators requires the determination of the following four parameters:

– oscillation frequency;

– condition for sustained oscillation;

– amplitude of oscillation;

– distortion factor of the generated signal.

It has already been shown that the oscillation frequency and the condition for sustained oscillation can be determined by analyzing the oscillator as a linear device and applying the Barkhausen condition. As for the amplitude and the distortion factor, it is practically impossible to predict them without resorting to the nonlinear theory of oscillators. Let us suppose that the oscillator operates under conditions of self-stabilization of the amplitude of the generated signal (dynamic equilibrium). To illustrate the problems related to the nonlinearity of the active element, let us consider the example of the Wien bridge oscillator, as schematically shown in Figure 4.6.

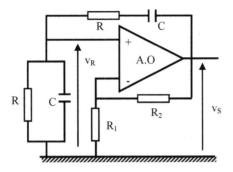

Figure 4.6. *Wien bridge oscillator*

Let us calculate separately the amplifier gain and the transfer function of the feedback circuit. Amplification (Figure 4.7(a)) is given by:

$$\frac{v_S}{v_e} = \frac{R_2 + R_1}{R_1} = A$$

It is worth recalling that in this case it is possible to separate the amplification and feedback circuits without taking into account the impedance balancing, given that the differential impedance of the operational amplifier is very high. It does not charge the feedback circuit.

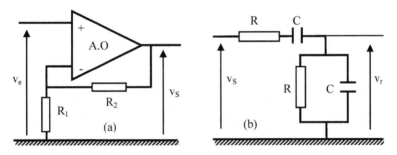

Figure 4.7. *a) Amplifier element of the Wien bridge oscillator.*
b) Feedback circuit of the Wien bridge oscillator

For the feedback circuit (Figure 4.7(b)), the following relation is obtained:

$$\frac{v_r}{v_S} = \frac{1}{3 + j(RC\omega - \dfrac{1}{RC\omega})} = \beta$$

Operating in the Laplace domain and considering:

$$\omega_0 = \frac{1}{RC} \quad \text{and} \quad p = j\omega$$

Grouping the amplifier gain and the transfer function of the feedback circuit leads to:

$$\frac{v_r}{v_e} = \frac{v_r}{v_s}\frac{v_s}{v_e} = \frac{A}{3 + \dfrac{p}{\omega_0} + \dfrac{\omega_0}{p}} = \frac{pA\omega_0}{p^2 + 3p\omega_0 + \omega_0^2}$$

$$(p^2 + 3p\omega_0 + \omega_0^2)v_r = (pA\omega_0)v_e$$

Writing this equation in terms of a differential equation leads to:

$$(\frac{d^2 v_r}{dt^2} + 3\omega_0\frac{dv_r}{dt} + \omega_0^2 v_r) = A\omega_0\frac{dv_e}{dt}$$

It can be noted that $v_r = v_e$ (see Figures 4.6 and 4.7(a)). Hence:

$$\frac{d^2 v_e}{dt^2} + (3 - A)\omega_0\frac{dv_e}{dt} + \omega_0^2 v_e = 0$$

It can be noted that there are three possibilities for voltage v_e and consequently for the output voltage (v_s) of the oscillator (Figure 4.8) depending on the value of gain A.

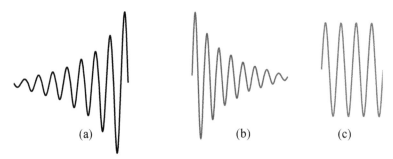

(a)   (b)   (c)

Figure 4.8. *Evolution of the output voltage depending on the gain value: a) A > 3; b) A < 3 and c) A = 3. For a color version of this figure, see www.iste.co.uk/haraoubia/nonlinear1.zip*

There is a nonlinear aspect to the amplifier gain A. The variation of output voltage "v_S" of the amplifier depends on its input voltage "v_e" (Figure 4.9). It is a function of the gain of the amplifier in question.

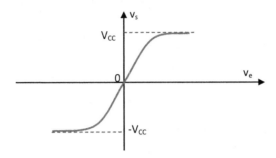

Figure 4.9. *Response curve of the amplifier*

When the oscillator starts operating (oscillator supply), the transients or the noise existing in the circuit allow the generation of a signal of frequency f_0 (frequency selected by the feedback circuit).

In order to initiate the generation of the output signal, it is absolutely necessary to have $A.\beta(f_0) > 1$.

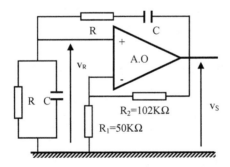

Figure 4.10. *Wien bridge oscillator with gain A slightly above 3*

To ensure the start of oscillations, a gain A that is real and slightly above 3 will be fixed (Figure 4.10).

$$A = \frac{R_1 + R_2}{R_1}$$

$$A = 3.04$$

$$f_0 = \frac{1}{2\pi RC}$$

The simulation of the response of this circuit is shown in Figure 4.11, in which output voltage v_S and voltage v_R returned by the feedback circuit to the non-inverting input of the operational amplifier are simultaneously represented. Imposing $A\beta(f_0) > 1$ at the beginning indeed shows the start of oscillations.

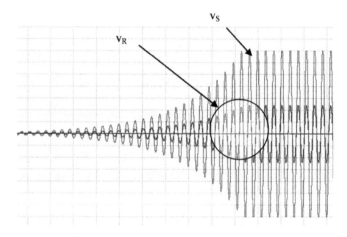

Figure 4.11. *Evolution of signal v_S at the output of Wien bridge oscillator and of voltage vr returned to the input by the feedback circuit with A = 3.04. For a color version of this figure, see www.iste.co.uk/haraoubia/nonlinear1.zip*

Output voltage increases exponentially until its amplitude reaches the value of the supply voltage. At that moment, it gets stabilized and the signal is no longer perfectly sinusoidal (Figure 4.12). Its peak is clipped around the supply voltage of the operational amplifier. Distortions are therefore evident in the output signal.

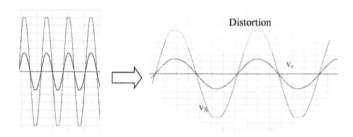

Figure 4.12. *Distortion problem when the loop gain is above 1. For a color version of this figure, see www.iste.co.uk/haraoubia/nonlinear1.zip*

The spectrum of the output signal contains not only the oscillation frequency (first harmonic f_0) to be generated but also a certain number of higher harmonics.

It can be shown experimentally that during the stabilization of the output signal amplitude (peak clipping), the gain of the first harmonic is no longer equal to 3, but below this value.

IN SUMMARY.–

1) An oscillator with a loop gain $A\beta < 1$ cannot operate, as oscillations cannot start.

2) An oscillator with a loop gain $A.\beta > 1$ cannot generate a purely sinusoidal oscillation, because the oscillation increases to signal peak clipping and the spectrum of the output signal contains several harmonics. When $A\beta > 1$, the generated signal is rather similar to a square-wave signal than to a sinusoidal signal.

3) The Barkhausen condition ($A\beta = 1$) is very difficult to obtain with satisfactory stability due to the accuracy of components and especially their derivatives.

4) Proper operation of an oscillator requires its loop gain to be slightly above 1 ($A\beta > 1$) for low-amplitude signals, to ensure the start of oscillations.

5) When the signal reaches too high an amplitude, the amplifier gain "A" must decrease in order to obtain a loop gain below 1 ($A\beta < 1$).

6) Amplitude stabilization requires having such a gain "A" at oscillation frequency that the loop gain is close to 1 ($A\beta \cong 1$): point of dynamic stability.

4.4. Stabilization of the amplitude of output voltage

As shown above, the oscillator is a nonlinear device, the loop gain of which depends on the amplitude of the signal to be generated (Figure 4.13).

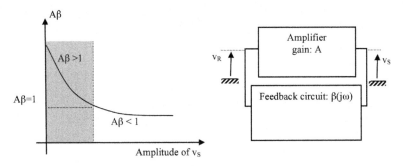

Figure 4.13. *Nonlinear variation of the loop gain of an oscillator*

In addition to a selective feedback circuit, generating a sinusoidal wave requires meeting the previously described conditions:

$A\beta > 1$ to trigger oscillations. For the present case, A is positive real (Wien bridge oscillator), and it should satisfy the relation: $A = A_1 > 3$;

$A\beta < 1$ for high amplitudes ($A = A_2 < 3$);

$A\beta = 1$ for the amplitude expected to be generated $A = A_0 = 3$.

Figure 4.14. shows an example of a circuit that could meet these three conditions.

A classic Wien bridge oscillator can be noted, to which nonlinear elements have been added in order to provide the nonlinear aspect of the oscillator and stabilize the amplitude of the output signal.

Certainly, some other methods can be used to stabilize the output signal at well-determined amplitudes.

In our opinion, Figure 4.14 illustrates the simplest didactic approach.

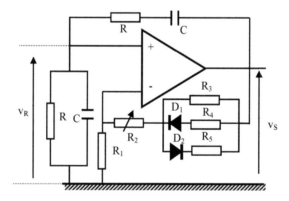

Figure 4.14. *Control of the amplitude of the output signal by the introduction of nonlinear elements in the feedback loop of the oscillator*

The signal that can be obtained is schematically shown in Figure 4.15. Initially, the amplifier gain is above 3, which triggers the oscillations. If the diode circuit is not activated, the output signal goes to saturation and is peak-clipped.

As soon as the diode circuit is activated, amplitude is stabilized at V_M and the signal is sinusoidal.

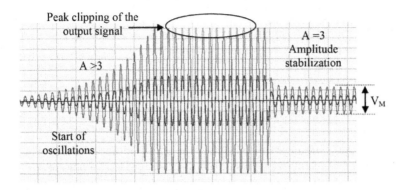

Figure 4.15. *Evolution of the output signal as a function of the amplifier gain. Stabilization of the amplitude by activation of diodes D_1 and D_2*

4.5. Amplitude of the output signal: first harmonic method

4.5.1. *Principle of the first harmonic method*

As already mentioned, oscillators are loop devices. Feedback is positive. Amplification contains a nonlinear element for the stabilization of the amplitude of the output signal (Figure 4.16). In an oscillator, voltage v_e is zero.

$$v_1 = v_r + v_e$$

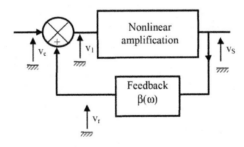

Figure 4.16. *Nonlinear amplification in an oscillator*

Voltage v_1 is sinusoidal, but voltage v_s may not be sinusoidal due to the nonlinearity introduced by amplification. However, v_s is periodical. Therefore, this voltage can be expanded in terms of Fourier series as:

$$v_1 = V_M \sin(2\pi f_0 t)$$

$$v_s = a_0 + \sum a_n \sin(2\pi n f_0 t + \phi)$$

where "a_0" is the continuous component of the signal at the amplifier output and "a_n" are the amplitudes of various harmonics that constitute the output signal.

The first harmonic method involves retaining only the first harmonic ($n = 1$) constituting the output signal "v_s". Consequently, the expression of the output signal is given by:

$$v_{S1H} = a_1 \sin(2\pi f_0 t + \varphi)$$

In an oscillator, this is explained by the fact that the feedback circuit is very selective and allows only the passage of the oscillation frequency "f_0".

4.5.2. *Study of amplitude stabilization*

In steady state, the nonlinearity of the amplifier is manifest. The amplifier output is influenced by the nonlinear aspect introduced by the amplifier limits. The output is no longer sinusoidal. To obtain a sinusoidal signal, the first harmonic must be extracted. This nonlinearity is modeled by an equivalent nonlinear gain denoted by A_{NL}, which is the ratio of the first harmonic of the output signal to the input signal of the amplifier. A_{NL} is the steady-state gain that should replace the gain A existing upon oscillator start.

The condition for sustained oscillation is then given (established steady state):

$$A_{NL}.\beta(j\omega) = 1: \text{Barkhausen condition.}$$

The solutions to this equation allow the determination of the oscillation frequency and of the amplitude of the generated signal.

A simple oscillator such as the Wien bridge oscillator will be used to illustrate the nonlinear effect in an oscillator, the choice of the first harmonic method and how the amplitude of the output signal in an oscillator can be stabilized.

It should be recalled that the operation of this Wien bridge oscillator is characterized by the following second-order differential equation:

$$\frac{d^2 v_S}{dt^2} + (3 - A_{NL})\omega_0 \frac{dv_S}{dt} + \omega_0^2 v_S = 0$$

When the feedback circuit is quite selective, a limit solution can be obtained at output:

$$v_{S1H} = a_1 \sin(\omega_0 t + \varphi) \quad \text{with: } \omega_0 = \frac{1}{RC}$$

This solution is obtained for:

$$3 - A_{NL} = 0.$$

The response curve (Figure 4.17(a)) of the amplifier can be identified with the curve described by the mathematical relation $y = \tanh(x)$ shown in Figure 4.17(b).

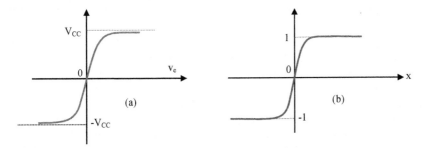

Figure 4.17. *Similarity between a) the response $v_s = f(v_e)$ of the amplifier related to Wien bridge oscillator and b) the plot of function $y = \tanh(x)$*

Given the above, the following relation can be written:

$$v_s = \alpha \tanh(\gamma v_e); \text{ with } \alpha = V_{CC} \text{ and } \gamma = (A/V_{CC})$$

This allows the deduction of the output voltage expression as a function of input voltage:

$$v_S = V_{CC} \cdot \tanh(\frac{A}{V_{CC}} v_e)$$

The expansion of function tanh(x) to the third degree yields:

$$v_S = V_{CC}.\tanh(\frac{A}{V_{CC}}v_e) = V_{CC}\left[\frac{A}{V_{CC}}v_e - \frac{1}{3}(\frac{A}{V_{CC}}v_e)^3\right]$$

The wave v_e is sinusoidal and can be written as:

$$v_e = V_M.\sin(\omega_0 t)$$

$$v_S = V_{CC}\left[\frac{A}{V_{CC}}V_M.\sin(\omega_0 t) - \frac{1}{3}(\frac{AV_M}{V_{CC}})^3\sin^3(\omega_0 t)\right]$$

$$\sin^3(\omega_0 t) = \frac{3}{4}\sin(\omega_0 t) - \frac{1}{4}\sin(3\omega_0 t) + ...$$

When the expansion is limited to the third degree, the following is obtained:

$$v_S = V_{CC}\left[\frac{A}{V_{CC}}V_M.\sin(\omega_0 t) - \frac{1}{3}(\frac{AV_M}{V_{CC}})^3\left[\frac{3}{4}\sin(\omega_0 t) - \frac{1}{4}\sin(3\omega_0 t)\right]\right]$$

When only the first harmonic is considered, the following can be written:

$$v_{S_{1H}} = V_{CC}\left[\frac{A}{V_{CC}}V_M.\sin(\omega_0 t) - \frac{1}{4}(\frac{AV_M}{V_{CC}})^3.\sin(\omega_0 t)\right]$$

The nonlinear gain is given by:

$$A_{NL} = A - \frac{A^3}{4}(\frac{V_M}{V_{CC}})^2 : \text{Approximate relation.}$$

This relation has been reached in an approximate manner, which evidences the nonlinear aspect of oscillators.

To obtain a sustained sinusoidal wave at output, the Barkhausen condition should be verified:

$$A_{NL}.\beta = 1 \Leftrightarrow A - \frac{A^3}{4}(\frac{V_M}{V_{CC}})^2 = 3$$

$$A_{NL}.\beta = 1 \Leftrightarrow (A-3) = V_{CC}.\frac{A^3}{4}(\frac{AV_M}{V_{CC}})^2$$

Based on this equation, peak amplitude V_M related to the first harmonic of the returned signal and, consequently, that of the output signal can be readily determined:

$$V_M = 2V_{CC}\sqrt{\frac{A-3}{A^3}}$$

where A is the gain required for the oscillations to start, which should imperatively be above 3.

$$A = 3+a; \ a > 0$$

$$V_M = 2V_{CC}\sqrt{\frac{(a)}{(3+a)^3}}$$

4.6. Exercises

Exercise 1

The task is to realize an oscillator based on the feedback circuit schematically shown in Figure E1.1.

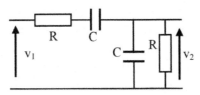

Figure E1.1.

1) Find the expression of the transfer function ($\beta = v_2/v_1$) of this circuit.

2) The active element used in the amplification has a real gain. It is built around an operational amplifier. Given these conditions, determine the overall electric diagram of the oscillator.

3) Find the oscillation frequency and the condition for sustained oscillation of this oscillator.

EXERCISE 2

Let us have an amplifier A of gain G and input resistance R_e: $R_e \longrightarrow \infty$. This amplifier is associated with a feedback loop formed of a passive quadripole B that introduces attenuation β (Figure E2.1).

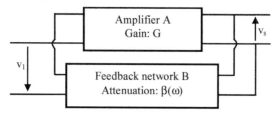

Figure E2.1.

1) Find the expression of ratio (v_s/v_1).

2) What condition should be met for the voltage v_s to maintain a finite value when $v_1 \rightarrow 0$?

3) What is to be said about this system when the previous condition is verified?

Let us consider the passive quadripole in Figure E2.2.

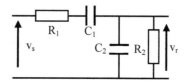

Figure E2.2.

4) Find the expression that defines the transfer function of this circuit. This quadripole occupies the position of the passive circuit designated by B in Figure E1.1 in order to realize an oscillator. Draw the diagram of this oscillator.

5) Find the oscillation frequency and the condition for sustained oscillation of this oscillator.

Numerical application: $R_1 = R_2 = 2.2 \ k\Omega$; $C_1 = C_2 = 0.22 \ \mu F$.

EXERCISE 3

Given the circuit in Figure E3.1, the task is to study it and make it operate as an oscillator.

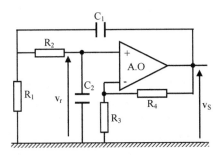

Figure E3.1.

1) Separate the amplification circuit and the feedback circuit, which sets the oscillation frequency, and draw the diagram of each of these circuits.

2) Find the expression of the transfer function of each of these two circuits: feedback circuit, denoted by β, and amplification circuit, denoted by A.

3) What form does the expression of β take if: $R_1 = R_2 = R$ and $C_1 = C_2 = C$?

4) Is it possible to apply the Barkhausen condition directly? Find the oscillation frequency of the output signal and the condition for sustaining these oscillations if $R_1 = R_2 = R$ and $C_1 = C_2 = C$.

5) Given these conditions, plot the voltages v_s and v_r on the same diagram. What conclusions can be drawn?

EXERCISE 4

An oscillator (Figure E4.1) is formed of an amplifier and a feedback circuit. The Bode plot of the feedback circuit is shown in Figure E4.2.

1) Assuming that the amplifier has a positive real gain, determine the oscillation frequency and the amplifier gain required to sustain this oscillation.

2) Considering now that the amplifier used has a negative real gain, answer the same questions formulated at point 1.

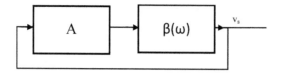

Figure E4.1.

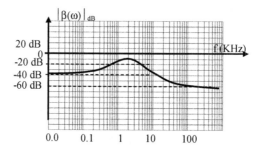

Figure E4.2(a). Bode plot (module)

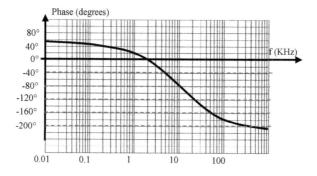

Figure E4.2(b). Bode plot (phase)

EXERCISE 5

The operational amplifier circuit shown in Figure 5.1(a) is the object of our study.

1) Find the expressions of ratios (v_1/v_e) and (v_2/v_1). What is the overall voltage gain of the circuit shown in Figure E5.1(a)?

2) What is the phase shift introduced between v_1 and v_e and then between v_2 and v_e?

3) Calculate the input resistance of the circuit shown in Figure E5.1(a).

4) The task is to realize an oscillator by connecting the circuits shown in Figures E5.1(a) and E5.1(b), as indicated. Find the oscillation frequency and the condition that R_1 and R_2 should satisfy for sustained oscillation.

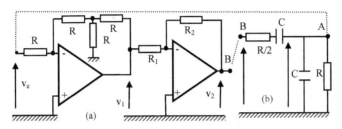

Figure E5.1.

EXERCISE 6

Let us consider the circuit shown in Figure E6.1. The operational amplifier is assumed ideal.

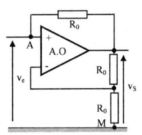

Figure E6.1.

1) Prove that the impedance seen between the points A and M is negative.

2) The diagram (Figure E6.1) is transformed in order to realize the oscillator schematically presented in Figure E6.2. Find the oscillation frequency and the condition that R should satisfy in order to sustain this oscillation. Draw a simplified diagram of this oscillator.

3) Draw the curve of the output voltage when $R = (R_0/2)$.

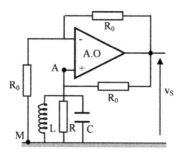

Figure E6.2.

EXERCISE 7

Let us consider the passive circuit shown in Figure E7.1.

1) Calculate the ratio (v_s/v_e)

2) It is our intention to use this circuit to build an oscillator. Suggest an oscillator circuit whose feedback is the passive device shown in Figure E7.1.

3) Find the oscillation frequency and the condition for sustained oscillation of the suggested oscillator.

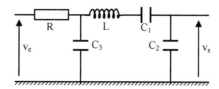

Figure E7.1.

EXERCISE 8

The circuit to be studied is schematically shown in Figure E8.1.

1) Calculate the ratio (v_s/v_e) putting it in the following form:

$$\frac{v_s}{v_e} = \frac{k}{\left[1 + j\frac{\omega}{\omega_0}(x-k) - \left[\frac{\omega}{\omega_0}\right]^2 \right]}$$

2) Find the expressions of x, k and ω_0.

3) Draw the variations of the curve $A_{dB} = 20.\log_{10}|v_s/v_e|$ and then deduce the condition that this circuit should meet in order to become an oscillator.

4) Let us now consider the passive circuit shown in Figure E8.2. Write its transfer function.

5) The passive circuit shown in Figure E8.2 is associated with an amplifier in order to obtain the circuit schematically shown in Figure E8.3. Find the relation between R_1 and R_2 that allows this device (Figure E8.3) to operate as an oscillator. Determine its oscillation frequency.

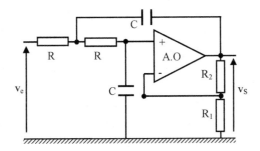

Figure E8.1.

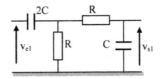

Figure E8.2.

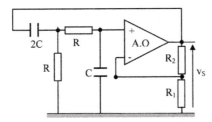

Figure E8.3.

EXERCISE 9

Let us consider the two circuits schematically shown in Figures E9.1 and E9.2.

1) Express the ratios $\dfrac{v_{e1}}{v_{s1}}$ and $\dfrac{v_{e2}}{v_{s2}}$

2) Since these two circuits are oscillators, calculate the oscillation frequency and find the condition for sustained oscillation for each of these two circuits.

3) What conclusion can be drawn concerning this type of oscillators?

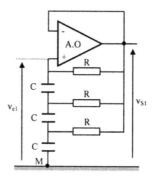

Figure E9.1.

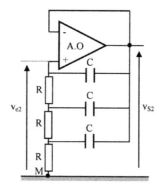

Figure E9.2.

EXERCISE 10

The task is to study the circuits represented in Figures E10.1, E10.2 and E10.3.

1) Calculate the ratio (v_s/v_e) for the circuit shown in Figure E10.1.

2) Determine the gain for the circuits shown in Figures E10.2 and E10.3.

3) Determine a diagram of an oscillator that uses the device shown in Figure E10.1 as the feedback circuit and the appropriate circuit chosen among those represented in Figures E10.2 and E10.3 as the amplification element.

4) Calculate the oscillation frequency and find the condition for sustained oscillation of the chosen device.

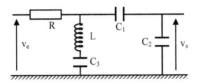

Figure E10.1.

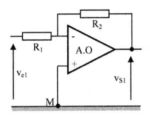

Figure E10.2.

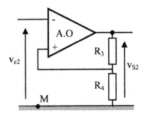

Figure E10.3.

Numerical application:

$C_1 = 2$ nF; $C_2 = 100$ nF; $C_3 = 200$ pF; $L = 100$ μH and $R_2 = 100$ kΩ.

EXERCISE 11

Let us consider the device shown in Figure E11.1. The blocks B_1 and B_2 are two passive quadripoles, the transfer functions of which are the quantities β_1 and β_2, respectively. A_1 is an ideal operational amplifier. Its open-loop gain is equal to A_0, the value of which is very high and therefore can be considered infinite.

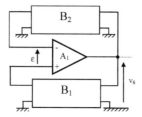

Figure E11.1.

1) Establish the relationship among A_0, β_1 and β_2.

The quadripoles B_1 and B_2 are replaced by their equivalent circuit, as indicated in Figure E11.2.

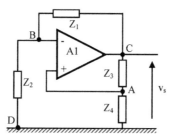

Figure E11.2.

2) Find the expressions that define β_1 and β_2.

3) Considering that the circuit shown in Figure E11.3 replaces the two feedback networks B_1 and B_2, establish the diagram of the oscillator circuit obtained and determine the oscillation frequency and the condition for sustained oscillation.

4) Let us now consider that the bridge represented by the circuit schematically shown in Figure E11.4 takes the place of the bridge formed by B_1 and B_2. Determine the diagram of the obtained circuit as well as the oscillation frequency of this circuit and the condition for sustained oscillation.

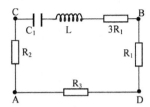

Figure E11.3.

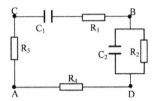

Figure E11.4.

EXERCISE 12

Let us consider the passive circuit schematically shown in Figure E12.1.

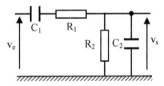

Figure E12.1.

1) Find the expression of the ratio v_s/v_e.

2) Represent the variations of the ratio (v_s/v_e) as a function of module and phase. What conclusions can be drawn?

3) Let us now study the low-frequency oscillator circuit of the circuit schematically shown in Figure E12.2.

Determine the oscillation frequency and the condition for sustained oscillation of this device.

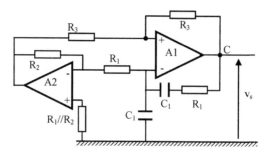

Figure E12.2.

EXERCISE 13

Let us consider the network shown in Figure E13.1.

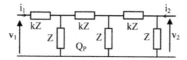

Figure E13.1.

1) Using Kirchhoff's second law, find the parameters Z_{ij} of the quadripole $\mathbf{Q_p}$ as a function of k and Z (k is an imaginary number).

This device serves for building an oscillator as schematically shown in Figure E13.2. The active quadripole $\mathbf{Q_A}$ is represented by its parameters Y_{ij}, which are real numbers (with $Y_{11} = 0$ and $Y_{12} = 0$).

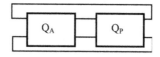

Figure E13.2.

The operating equation of the oscillator is given by:

$$Z_{21}(Y_{21} - Y_{12}) + Y_{11}Z_{22} + Y_{22}Z_{11} + \Delta Y.\Delta Z + 1 = 0$$

with $\Delta Z = Z_{11}.Z_{22} - Z_{12}.Z_{21}$ and $\Delta Y = Y_{11}.Y_{22} - Y_{12}.Y_{21}$

2) Let us consider $Z = R$; $kZ = -jaZ$ ($a \neq 0$) is the impedance related to a capacitor C. Under these conditions, find the expression of k and a.

3) Write the impedances Z_{ij} as a function of "a" and of R. Write the operating equation of the oscillator as a function of "a", R and Y_{ij}.

4) Assuming that Y_{ij} are real parameters, find the oscillation frequency of this circuit. Determine its expression when $Y_{22}.R \gg 1$.

EXERCISE 14

Let us consider the oscillator shown in Figure E14.1. The capacitor C_0 is considered a short circuit at working frequency. Resistance R_E is very high compared to h_{11}.

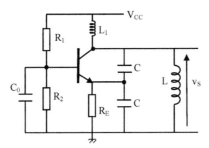

Figure E14.1.

The operating equation of the oscillator is:

$$-Y_{21}(h_{21} - h_{12}) + Y_{11} + h_{22} + \Delta h Y_{22} + \Delta Y h_{11} = 0$$

The active element is represented by parameters h_{ij}, and the passive element is represented by parameters Y_{ij}. Let us assume that $h_{12} \cong 0$ and $h_{22} \cong 0$.

1) What is the type of oscillator studied and what is the role of self-inductance L_1? What is the role of capacitor C_0?

2) What is the configuration in which the transistor is connected (common emitter, common base or common collector)? Justify your answer.

3) Find the dynamic equivalent diagram of this circuit evidencing the amplifier chain and the feedback loop. Let us assume that $h_{12} = 0$ and $h_{22} = 0$.

4) Calculate the parameters Y_{ij} of the passive quadripole.

5) Write the operating equation of this oscillator and determine the oscillation frequency.

EXERCISE 15

Let us consider the diagram in Figure E15.1, which is the equivalent diagram of a quartz crystal.

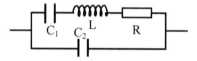

Figure E15.1.

1) Find the admittance of the quartz crystal.

2) Find the condition that allows us to write this admittance in the following form:

$$Y = j2\pi F(C_1 + C_2)\frac{1-(F/f_p)^2}{1-(F/f_S)^2}$$

3) Find the expression of frequency f_p as a function of frequency f_s as well as the values of f_s and f_p. Given: $R = 10\ \Omega$, $C_1 = 1$ fF, $C_2 = 5$pF and $L = 250$ mH.

4) The quartz crystal operates at a frequency F_0, with $f_s < F_0 < f_p$. Given these conditions, find the expression of the reactance presented by the quartz crystal, indicate its sign and deduce its nature.

5) The quartz crystal is used in oscillation; its quality factor (series resonance) is given by $Q = (1/2\pi RC_1 F_0)$. What can be deduced from this, given that a good LC circuit has a quality factor $Q_{LC} = 100$?

EXERCISE 16

Let us consider the diagram of the sinusoidal oscillator in Figure E16.1. The transistor quadripole is represented by its hybrid parameters, and the passive feedback quadripole is represented by its impedance parameters.

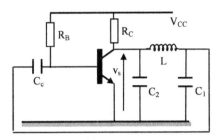

Figure E16.1.

Without introducing the influence of R_C and R_B, the operating equation of the oscillator is given by $Z_{21}(h_{21} - h_{12}) + Z_{22} + h_{11} + \Delta h Z_{11} + \Delta Z h_{22} = 0$.

1) Find the equivalent diagram of this circuit. Explain the role of capacitor C_c. ($h_{12} = 0$ and $h_{22} = 0$ for the transistor).

2) An equivalent diagram of this oscillator is shown in Figure E16.2. Identify the elements of Figure E16.2 with the elements of the equivalent diagram you have proposed.

3) As indicated, the transistor is represented by its hybrid parameters and the passive quadripole by its Z parameters.

3.1) Find the Z_{ij} parameters of the passive quadripole as a function of Z_1, Z_2 and Z, then of C_1, C_2, L and ω.

3.2) Write the operating equation in the presence of R_B and R_c.

3.3) If the influence of R_B and of R_C is now neglected (the values of resistances R_C and R_B are chosen sufficiently high), then find the oscillation frequency under these conditions.

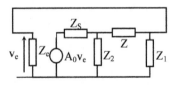

Figure E16.2.

EXERCISE 17

Let us consider the circuit shown in Figure E17.1, related to the equivalent diagram of a quartz crystal:

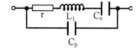

Figure E17.1.

1) Calculate the impedance of this circuit and show that it can be written in the following form:

$$Z = \frac{1 + jC_s\omega(r + jL\omega)}{j(C_s + C_p)\omega - C_sC_p\omega^2(r + jL\omega)} \quad \text{with} \quad C_p \gg C_s$$

2) If resistance r is neglected (r = 0), then find the simplified expression of the impedance of quartz crystal as well as its value for a continuous signal.

3) Prove that the simplified expression can be written in the following form:

$$Z = \frac{1}{jC_{eq}\omega} \left[\frac{1 - \left[\dfrac{\omega}{\omega_1}\right]^2}{1 - \left[\dfrac{\omega}{\omega_2}\right]^2} \right]$$

4) Find the expressions of C_{eq}, ω_1 and ω_2. Prove that $\omega_1 < \omega_2$. What is the significance of ω_1 and ω_2?

5) Study the variations in terms of module and phase of the impedance Z as a function of ω and indicate the nature of Z in the various ranges separated by ω_1 and ω_2.

EXERCISE 18[1]

Let us consider the circuit shown in Figure E18.1, which is intended to be used in an oscillator circuit, where "m" is a positive constant.

1) Calculate the ratio (v_2/v_1) using Kirchhoff's second law.

2) Plot the evolution (v_2/v_1) in terms of module and phase as a function of frequency. Indicate the sign of (v_2/v_1) when its imaginary part is zero.

---

1 No solution has been provided for this exercise in order to allow the reader to test themselves.

3) This circuit is used in the realization of an oscillator that uses an operational amplifier or a bipolar transistor as amplifier element. Draw the diagram of this oscillator for each type of active component. Find the oscillation frequency and the gain of the amplifier that has been proposed for the two types of active components.

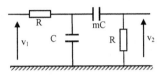

Figure E18.1.

EXERCISE 19[2]

Let us consider the circuit shown in Figure E19.1 and the block diagram shown in Figure E19.2. The operational amplifier is ideal and operates under linear conditions.

1) Identify the amplifier and feedback blocks by drawing them separately, while indicating the position of voltages schematically shown in Figure E19.2 in the diagram shown in Figure E19.1.

2) Calculate the ratios (v_2/v_1) and (v_3/v_2) and find the relationship between R_1 and R_2 so that this circuit is an oscillator.

3) Under these conditions, find the oscillation frequency.

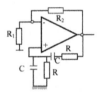

Figure E19.1.

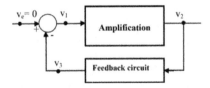

Figure E19.2.

2 No solution has been provided for this exercise in order to allow the reader to test themselves.

4.7. Solutions to exercises

SOLUTION TO EXERCISE 1

1) Expression of the transfer function of the circuit shown in Figure E1.2

$$\beta = \frac{v_2}{v_1} = \frac{Z_P}{Z_P + Z_S} = \frac{1}{1 + \dfrac{Z_S}{Z_P}}$$

$$\beta = \frac{RC\omega}{3RC\omega + j((RC\omega)^2) - 1)}$$

2) Diagram of the oscillator

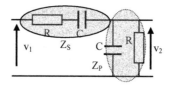

Figure E1.2.

The gain A of the amplifier is real. The condition for operation is $A.\beta = 1$. Consequently, β must also be real (the imaginary part is zero). Under these conditions, β is > 0 and equal to (1/3). The gain A must also be real and equal to 3.

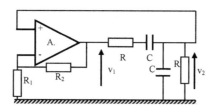

Figure E1.3. *Diagram of the designed oscillator*

For the amplifier element and for the oscillator, this leads to the diagram in Figure E1.3.

3) Oscillation frequency and condition for sustained oscillation

3.1) Oscillation frequency

$$\text{Im}(\beta) = 0 \Leftrightarrow (RC\omega_0)^2 - 1 = 0 \ ; \ \omega_0 = \frac{1}{RC} \ \ f_0 = \frac{1}{2\pi RC}$$

3.2) Condition for sustained oscillation

$$A.\beta\big|_{\omega=\omega_0} = 1 = \frac{R_2+R_1}{R_1}.\beta$$

$$\beta = \frac{1}{3}$$

$A = 3$ and $R_2 = 2R_1$

SOLUTION TO EXERCISE 2

1) Expression of the ratio (v_s/v_1)

$$\frac{v_s}{v_1} = \frac{G}{1-G\beta}$$

2) Condition that allows v_s to maintain a finite value when $v_1 \to 0$

The following condition should be met: $G.\beta = 1$.

3) Under these conditions, it can be stated that the system is unstable. It will therefore oscillate.

4) Expression of the transfer function and diagram of the oscillator

4.1) Expression of the transfer function

$$\beta = \frac{1}{1+\dfrac{R_1}{R_2}+\dfrac{C_2}{C_1}+j(R_1C_2\omega-\dfrac{1}{R_2C_1\omega})}$$

4.2) Diagram of the oscillator

The diagram of the oscillator is shown in Figure E2.3.

5) Oscillation frequency and condition for sustained oscillation

$$f_0 = \frac{1}{2\pi\sqrt{R_1R_2C_1C_2}} ; \quad G = 1 + [(R_1C_1+R_2C_2)/R_2C_1]$$

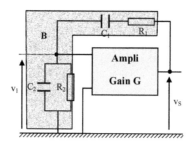

Figure E2.3.

Numerical applications: $R_1 = R_2 = 2.2 \text{ k}\Omega$; $C_1 = C_2 = 0.22 \text{ μF}$

$f_0 = 329 \text{ Hz}$; $G = 3$.

SOLUTION TO EXERCISE 3

1) Diagrams of the amplifier and feedback circuits

It can be readily noted that the amplifier circuit is represented by a non-inverting amplifier with the operational amplifier schematically shown in Figure E3.2(a). The feedback circuit is shown in Figure E3.2(b).

2) Expression of the transfer function of the two circuits

2.1) Amplifier circuit

$$A = \frac{v_s}{v_r} = \frac{R_4 + R_3}{R_3}$$

2.2) Feedback circuit

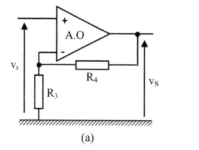

(a)

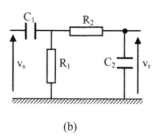

(b)

Figure E3.2.

In order to find the transfer function of the feedback circuit shown in Figure E3. 2(b), Kirchhoff's second law is used.

$$v_s = (Z_1 + R_1)i_1 - R_1 i_2 ; \quad Z_1 = (1/jC_1\omega)$$

$$0 = -R_1 i_1 + (R_1 + R_2 + Z_2)i_2; \quad Z_2 = (1/jC_2\omega)$$

$$v_r = Z_2 i_2$$

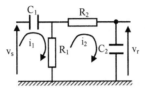

Then, i_1 is expressed as a function of v_r and the relation connecting v_r and v_s is deduced:

$$\beta = \frac{v_r}{v_s} = \frac{R_1 Z_2}{Z_1 Z_2 + R_1 R_2 + Z_1 R_1 + Z_1 R_2 + R_1 Z_2}$$

Replacing Z_1 and Z_2 by their equivalent expressions then leads to:

$$\beta = \frac{1}{j(R_2 C_2 \omega - \dfrac{1}{R_1 C_1 \omega}) + (1 + \dfrac{C_2}{C_1}) + \dfrac{R_2}{R_1}\dfrac{C_2}{C_1}}$$

3) Expression of β when $R_1 = R_2 = R$ and $C_1 = C_2 = C$

$$\beta = \frac{1}{3 + j(RC\omega - \dfrac{1}{RC\omega})}$$

4) It is possible to apply the Barkhausen condition directly. Oscillation frequency and condition for sustained oscillation: $R_1 = R_2 = R$ and $C_1 = C_2 = C$

It should be noted that the amplifier exhibits a very high input resistance. It is practically equal to the differential resistance of the operational amplifier, which is generally very high. Therefore, the amplifier does not load the feedback circuit, and the Barkhausen condition $A.\beta=1$ can be applied directly without seeking to reestablish the impedance balance.

4.1) Oscillation frequency

The gain provided by the amplifier is real, at oscillation frequency, and so should be the transfer function of the feedback circuit. Hence:

$$RC\omega_0 - \frac{1}{RC\omega_0} = 0 ; \qquad F_0 = \frac{1}{2\pi RC}$$

4.2) Condition for sustained oscillation

$$A.\beta(\omega_0) = 1 \Rightarrow \frac{R_3 + R_4}{R_3} \cdot \frac{1}{3} = 1 \Rightarrow R_4 = 2R_3$$

5) Representation of voltages v_s and v_r and conclusion

It can be noted that voltages v_s and v_r are in phase. This is in agreement with the theoretical aspect, since the feedback circuit introduces no phase shift between its input and output at oscillation frequency (β is positive real and equal to 1/3 at the oscillation frequency).

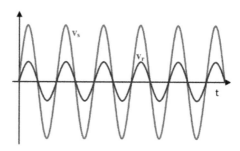

(For a color version of this figure, see www.iste.co.uk/haraoubia/nonlinear1.zip)

Voltage v_r is attenuated by a ratio of 1/3 with respect to v_s, which is utterly predictable, since at oscillation frequency, $\beta(\omega_0) = 1/3$.

SOLUTION TO EXERCISE 4

1) The amplifier has a positive gain. Determine the oscillation frequency and the amplifier gain in order to sustain this oscillation.

The amplifier has a real gain.

According to the Barkhausen condition:

$A.\beta(\omega) = 1$

A positive real $\Rightarrow Arg(\beta(\omega)) = \varphi = 0$.

The analysis of the curves (see Figure E4.3) relative to Bode plots evidences zero phase for a frequency equal to 2 kHz.

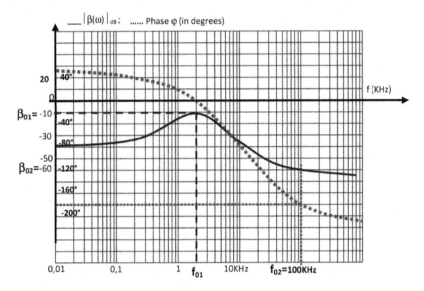

Figure E4.3. *Bode plot and determination of the characteristics of the oscillator. For a color version of this figure, see www.iste.co.uk/haraoubia/nonlinear1.zip*

Let us see what the value of β corresponding to this frequency is.

It can then be deduced: $f_{01} = 2$ kHz $\Rightarrow \beta(\omega_{01})_{dB} = -10$ dB;

$$\beta(\omega_{01})_{dB} = 20.\log(\beta(\omega_{01}); \ \beta(\omega_{01}) = 10^{\frac{\beta_{dB}}{20}}$$

$$\beta(\omega_{01}) = 10^{-10/20} = 0.316.$$

This yields: $A.\beta(\omega_{01}) = 1 \Rightarrow A = 3.16$.

2) Oscillation frequency and amplifier gain in order to sustain this oscillation. Case when the amplifier used has a negative real gain.

The previous approach is also applicable here, but now A is negative real. In order to meet the Barkhausen condition $A.\beta(\omega) = 1$, when the gain A is negative real, the argument of β at oscillation frequency should be equal to π radians (180°).

A negative real $\Rightarrow Arg(\beta(\omega_{02})) = \pi$;

As a result of the analysis of curves relative to the Bode plot, the following can be deduced:

$Arg(\beta(\omega_{02})) = \pi \Rightarrow f_{02} = 100$ kHz

Once the frequency has been determined, it is sufficient to note the correspondence between this frequency and the value of β in terms of module. This yields: $\beta(\omega_{02})_{dB} = -60$ dB $\Rightarrow A = 1,000$.

SOLUTION TO EXERCISE 5

1) Expressions of ratios (v_1/v_e) and (v_2/v_e) and overall voltage gain of the circuit shown in Figure E5.2

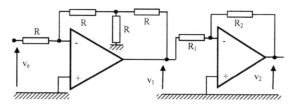

Figure E5.2.

The following results are obtained after calculation:

$$\frac{v_1}{v_e} = -3 \text{ and } \frac{v_2}{v_1} = -\frac{R_2}{R_1}$$

2) What phase shift is introduced between v_1 and v_e and then between v_2 and v_e?

The phase shift between v_1 and v_e is equal to π due to the negative sign existing in the relation between v_1 and v_e. The same is applicable to the phase shift between v_2 and v_1.

$$Arg(v_1/v_e) = Arg(v_2/v_1) = \pi + 2k\pi$$

The overall phase shift between v_2 and v_e is zero, due to the double phase shift of π each time.

$\text{Arg}(v_2/v_e) = 0 + 2k\pi$

3) Input resistance of the circuit shown in Figure E5.2

$R_e = R$

4) Calculation of the oscillation frequency and the condition for sustained oscillation

It is sufficient to calculate the gain A of the amplifier and the attenuation β of the feedback circuit and to apply the Barkhausen condition $A.\beta=1$.

In order to calculate the transfer function of the feedback circuit, impedance balance should be restored (the input resistance of the amplifier ($R_e = R$) is connected in parallel with resistance R of the feedback circuit in order to achieve an overall resistance $R/2$).

In the first stage, the calculation yields a relation that defines the overall gain brought by the amplifier: $A = 3(R_2/R_1)$.

The calculation of β relies on the diagram of the feedback circuit shown in Figure E5.3 (after restoring impedance balance):

$$\beta = \frac{v_e}{v_2} = \frac{R}{3R + j(\dfrac{R^2 C\omega}{2} - \dfrac{2}{C\omega})}$$

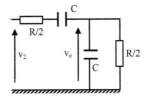

Figure E5.3.

In order to calculate the oscillation frequency, it is sufficient to cancel out the imaginary part of the transfer function of the feedback circuit, since the amplifier gain is positive.

$$\frac{R^2 C \omega_0}{2} - \frac{2}{C \omega_0} = 0 \Rightarrow F_0 = \frac{1}{\pi RC} \quad \text{(oscillation frequency)}$$

$$\beta(\omega_0) = \frac{1}{3} \quad A.\beta(\omega_0) = 1 \Leftrightarrow 3\frac{R_2}{R_1} \times \frac{1}{3} = 1 \Rightarrow R_2 = R_1$$

(condition for sustained oscillation).

SOLUTION TO EXERCISE 6

1) Impedance (Z_{AM}) seen between points A and M

$$Z_{AM} = \frac{V_e}{i_e} = -R_0$$

2) Oscillation frequency and condition that R should meet in order to sustain this oscillation, and simplified diagram of this oscillator

2.1) Oscillation frequency: $f_0 = \dfrac{1}{2\pi\sqrt{LC}}$

2.2) Condition on R for sustained oscillation: $R = R_0$.

2.3) Simplified diagram (see Figure E6.3) of the studied oscillator (when $R = R_0$)

$$f_0 = \frac{1}{2\pi\sqrt{LC}}$$

3) Curve of the output voltage v_S (see Figure E6.4) when $R = (R_0/2)$

Figure E6.3.

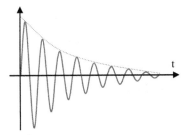

Figure E6.4. *The output wave is damped*

The output wave is damped, as the losses related to the presence of $R = (R_0/2)$ are only partially compensated.

SOLUTION TO EXERCISE 7

1) Ratio (v_s/v_e):

$$\frac{v_s}{v_e} = \frac{\dfrac{-1}{C_2 C_3 \omega^2}}{jR\left[L\omega - \dfrac{1}{\omega}\left[\dfrac{1}{C_1} + \dfrac{1}{C_2} + \dfrac{1}{C_3}\right]\right] + \dfrac{1}{C_3 \omega}\left[L\omega - \dfrac{1}{\omega}\left[\dfrac{1}{C_1} + \dfrac{1}{C_2}\right]\right]}$$

2) Proposed oscillator circuit

It should be noted that when the imaginary part of the transfer function of the feedback circuit (see the above ratio v_s/v_e) is cancelled out, this transfer function is negative.

The feedback circuit introduces a phase rotation equal to π. A choice should therefore be made for an amplifier that can also introduce an additional phase rotation equal to π in order to meet the condition for sustained oscillation (Barkhausen condition). The use of an operational amplifier connected as inverter is an appropriate solution to the problem posed.

The diagram of the oscillator is shown in Figure E7.2, with $R_1 \gg (1/C_2\omega)$

3) Oscillation frequency and condition for sustained oscillation

3.1) Oscillation frequency

$$f_0 = \frac{1}{2\pi}\sqrt{\frac{1}{L}\left[\frac{1}{C_1} + \frac{1}{C_2} + \frac{1}{C_3}\right]}$$

3.2) Condition for sustained oscillation

$$\frac{C_3}{C_2} = \frac{R_1}{R_2}$$

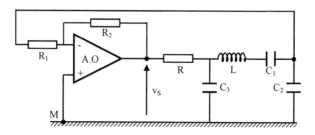

Figure E7.2. *Diagram of the oscillator*

SOLUTION TO EXERCISE 8

1) Ratio (v_s/v_e)

$$\frac{v_s}{v_e} = \frac{\dfrac{R_1 + R_2}{R_1}}{\left[1 + jRC\omega\left[3 - \dfrac{R_1 + R_2}{R_1}\right]\right] - (RC\omega)^2}$$

$$\frac{v_s}{v_e} = \frac{k}{1 + j\dfrac{\omega}{\omega_0}(x - k) - [\dfrac{\omega}{\omega_0}]^2}$$

2) Expressions of k, x and ω_0

  $k = (R_1 + R_2)/R_1$; $x = 3$ and $\omega_0 = (1/RC)$.

3) Variation of A_{dB}

The variation of $A = (v_s/v_e)$ expressed in decibels is shown in Figure E8.4.

The circuit can become an oscillator provided that $(v_s/v_e) \to \infty$ (and consequently the ratio $A = (v_s/v_e)$).

For this purpose, it is sufficient to find a gain k related to the amplifier in order to get $(v_s/v_e) \to \infty$.

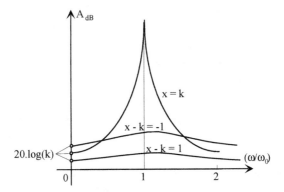

Figure E8.4. *Variation of the ratio (v_s/v_e) as a function of frequency*

It can be readily noted that this happens when $k = x = 3$. The circuit oscillates at the frequency $f_0 = (\omega_0/2\pi)$.

4) Transfer function of the passive circuit:

$$\frac{v_{s1}}{v_{e1}} = \frac{1}{2 + j(RC\omega - \dfrac{1}{2RC\omega})}$$

5) Relation between R_1 and R_2 so that the system oscillates and oscillation frequency:

$(R_1 + R_2)/R_1 = 2$;

$$R_2 = R_1; \quad f_0 = \frac{1}{\pi RC\sqrt{8}}$$

SOLUTION TO EXERCISE 9

1) Expression of ratios (v_{e1}/v_{s1}) and (v_{e2}/v_{s2})

1.1) Ratio (v_{e1}/v_{s1})

$$A.\beta = \frac{v_{e1}}{v_{s1}} = \frac{1}{1 + \dfrac{a^3}{j(1 - 6a^2) - 5a}}$$

where a = RCω

1.2) Ratio (v_{e2}/v_{s2})

$$\beta = \frac{v_{e2}}{v_{s2}} = \frac{1}{1 + \dfrac{-1}{ja(a^2-6)+5a^2}}$$

where a = RCω

2) Oscillation frequency and condition for sustained oscillation

2.1) Case of the circuit shown in Figure E9.1

$$f_0 = \frac{1}{2\pi RC\sqrt{6}}; \qquad A.\beta(\omega=\omega_0) = \frac{30}{29}$$

2.2) Case of the circuit shown in Figure E9.2

$$f_0 = \frac{\sqrt{6}}{2\pi RC}; \qquad A.\beta(\omega=\omega_0) = \frac{30}{29}$$

3) Conclusion to be drawn

It can be noted that it is possible to realize oscillators using only amplifiers with unity gain. Indeed, in both studied cases and according to the proposed configurations, the passive feedback circuit brings a voltage gain that is slightly above 1.

The choice of an amplifier with gain equal to unity having a high input resistance (in case of an operational amplifier connected as voltage follower) and connected according to Figure E9.1 or E9.2 enables the realization of an oscillator circuit.

SOLUTION TO EXERCISE 10

1) Expression of ratio (v_s/v_e)

$$\frac{v_s}{v_e} = \frac{\dfrac{1}{C_2\omega}\left[L\omega - \dfrac{1}{C_3\omega}\right]}{jR\left[L\omega - \dfrac{1}{\omega}\left[\dfrac{1}{C_1}+\dfrac{1}{C_2}+\dfrac{1}{C_3}\right]\right] + \left[\dfrac{1}{C_1\omega}+\dfrac{1}{C_2\omega}\right]\left[L\omega - \dfrac{1}{C_3\omega}\right]}$$

2) Gain of the circuits schematically represented in Figures E10.2 and E10.3

2.1) Case of the diagram shown in Figure E10.2

$$G_1 = \frac{v_{s1}}{v_{e1}} = \frac{-R_2}{R_1}$$

2.2) Case of the diagram shown in Figure E10.3

$$G_2 = \frac{v_{s2}}{v_{e2}} = \frac{R_3 + R_4}{R_4}$$

3) Diagram of the oscillator

It can be noted that the two amplifier circuits introduce a real gain. However, one introduces a phase rotation equal to π (the one shown in Figure E10.2). On the contrary, the circuit shown in Figure E10.3 introduces zero phase rotation. The feedback circuit has a complex transfer function. When the imaginary part of this function is cancelled out, it can be noted that it becomes positive real. In conclusion, the adequate amplifier circuit for realizing the oscillator in question is shown in Figure E10.3. The oscillator that can be realized is schematically shown in Figure E10.4.

4) Oscillation frequency and condition for sustained oscillation

4.1) Oscillation frequency

$$f_0 = \frac{1}{2\pi\sqrt{LC_{eq}}} \quad with \quad \frac{1}{C_{eq}} = \frac{1}{C_1} + \frac{1}{C_2} + \frac{1}{C_3}$$

4.2) Condition for sustained oscillation

$$\frac{R_4 + R_3}{R_4} = \frac{C_1 + C_2}{C_1}$$

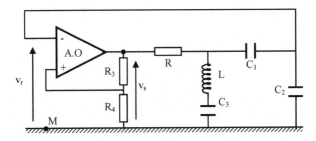

Figure E10.4. *Oscillator circuit*

Numerical application: $f_0 = 1.18$ MHz; $R_3 = 50R_4$.

SOLUTION TO EXERCISE 11

1) Relationship among A_0, β_1 and β_2

$A_0(\beta_1 - \beta_2) = 1$

2) Expressions that define β_1 and β_2

$$\beta_1 = \frac{Z_4}{Z_3 + Z_4}; \qquad \beta_2 = \frac{Z_2}{Z_1 + Z_2}$$

3) Diagram of the oscillator when the bridge schematically shown in Figure E11.3 is used; oscillation frequency and condition for sustained oscillation

3.1) Diagram of the oscillator (see Figure E11.5)

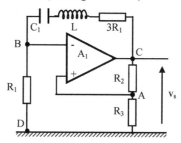

Figure E11.5. *Oscillator using as a feedback circuit the elements of the bridge shown schematically in Figure E11.3*

3.2) Oscillation frequency:

$$f_0 = \frac{1}{2\pi\sqrt{LC_1}},$$

3.3) Condition for sustained oscillation:

$R_2 = 3R_3$

4) Diagram of the oscillator when the bridge schematically shown in Figure E11.4 is used, and oscillation frequency and condition for sustained oscillation

4.1) Diagram of the oscillator (see Figure E11.6)

4.2) Oscillation frequency: $f_0 = \dfrac{1}{2\pi\sqrt{R_1 C_1 R_2 C_2}}$

4.3) Condition for sustained oscillation: $\dfrac{R_3 + R_4}{R_4} = 1 + \dfrac{R_1 C_1 + R_2 C_2}{R_2 C_1}$

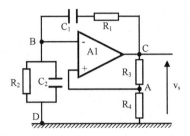

Figure E11.6. *Oscillator using as a feedback circuit the elements of the bridge shown schematically in Figure E11.4*

SOLUTION TO EXERCISE 12

1) Expression of the ratio v_S/v_e

$$\frac{v_S}{v_e} = \beta = \frac{R_2 C_1 \omega}{\omega(R_1 C_1 + R_2 C_1 + R_2 C_2) + j(R_1 C_1 R_2 C_2 \omega^2 - 1)}$$

2) Representation of the module and phase of the ratio (v_S/v_e)

2.1) Module:

$$\left|\frac{v_S}{v_e}\right| = |\beta| = \frac{R_2 C_1 \omega}{\sqrt{\left[\omega(R_1 C_1 + R_2 C_1 + R_2 C_2)\right]^2 + \left[(R_1 C_1 R_2 C_2 \omega^2 - 1)\right]^2}}$$

2.2) Phase:

$$\phi = -\mathrm{Arctg}\,\frac{R_1 C_1 R_2 C_2 \omega^2 - 1}{\omega(R_1 C_1 + R_2 C_1 + R_2 C_2)}$$

The variations of the module of β and its phase φ are shown in Figure E12.3.

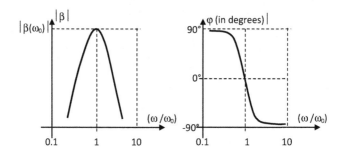

Figure E12.3. *Variation of the ratio (v_s/v_e) in module and in phase as a function of frequency*

It can be noted that the studied passive circuit can be used as a feedback element in an oscillator. This passive circuit allows the selection of frequency $f_0 = (\omega_0/2\pi)$. At this frequency, it can be noted that this circuit introduces a certain attenuation $\beta(\omega_0)$ and zero phase shift.

3) Oscillation frequency and condition for sustained oscillation

3.1) Oscillation frequency

$$f_0 = \frac{1}{2\pi R_1 C_1}$$

3.2) Condition for sustained oscillation

$$R_2 = 3R_1$$

SOLUTION TO EXERCISE 13

1) Parameters Z_{ij} of quadripole Q_p as a function of k and Z; applying Kirchhoff's second law to the circuit schematically shown below leads to the following equations:

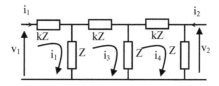

$$v_1 = kZi_1 + Z(i_1 - i_3) \tag{4.1}$$

$$0 = Z(i_3 - i_1) + kZi_3 + Z(i_3 - i_4) \tag{4.2}$$

$$0 = Z(i_4 - i_3) + kZi_4 + Z(i_4 + i_2) \tag{4.3}$$

$$v_2 = Z(i_4 + i_2) \tag{4.4}$$

Here, i_4 is determined as a function of i_3 and i_2 from equation [4.3]. It is then inserted in equation [4.2] in order to determine i_3 as a function of i_1 and i_2. Then, i_3 is replaced in equation [4.1], which yields v_1 as a function of i_1 and i_2. This allows the determination of Z_{11} and Z_{12} of quadripole Q_p. This yields:

$$v_1 = Z \frac{(1 + 6k + 5k^2 + k^3)}{3 + 4k + k^2} i_1 + \frac{Z}{3 + 4k + k^2} i_2$$

$$Z_{11} = Z \frac{(1 + 6k + 5k^2 + k^3)}{3 + 4k + k^2} \quad \text{and} \quad Z_{12} = \frac{Z}{3 + 4k + k^2}$$

Equation [4.4] is used in order to find Z_{21} and Z_{22}. i_4 is replaced by its equivalent expression in i_3 and i_2. Then, i_3 is replaced by its equivalent expression in i_1 and i_2. This leads to the expression of v_2 as a function of i_1 and i_2.

$$v_2 = \frac{Z}{3 + 4k + k^2} i_1 + \frac{(1 + 3k + k^2)}{3 + 4k + k^2} i_2$$

$$Z_{21} = \frac{Z}{3 + 4k + k^2} \quad \text{and} \quad Z_{22} = Z \frac{(1 + 3k + k^2)}{3 + 4k + k^2}$$

2) Expression of "k" and "a"

Let $Z = R$ and $kZ = -jaZ$ be the impedance related to a capacitor C.

$Z = R; kZ = (1/jC\omega)$:

$$\frac{kZ}{Z} = k = \frac{1}{jRC\omega} = \frac{-jaZ}{Z}; \quad k = \frac{-j}{RC\omega} \quad \text{and} \quad a = \frac{1}{RC\omega}$$

3) Expression of impedances Z_{ij} as a function of "a" and R, and operating equation of the oscillator as a function of "a", R and Y_{ij}

3.1) Expression of impedances Z_{ij}

$k = -ja;$

$$Z_{11} = R\frac{(1 - j6a - 5a^2 + ja^3)}{3 - 4ja - a^2} \quad \text{and} \quad Z_{12} = Z_{21} = \frac{R}{3 - 4ja - a^2}$$

$$Z_{22} = Z\frac{(1 - 3ja - a^2)}{3 + 4k + k^2} = R\frac{(1 - 3ja - a^2)}{3 - 4ja - a^2}$$

3.2) Operating equation

$$Z_{21}(Y_{21} - Y_{12}) + Y_{11}Z_{22} + Y_{22}Z_{11} + \Delta Y.\Delta Z + 1 = 0;$$

According to the given data: $Y_{12} = 0$ and $Y_{11} = 0 \Rightarrow \Delta Y = 0$

$$Z_{21}Y_{21} + Y_{22}Z_{11} + 1 = 0$$

$$\frac{R}{3 - 4ja - a^2}Y_{21} + Y_{22}.\frac{R.(1 - j6a - 5a^2 + ja^3)}{3 - 4ja - a^2} + \frac{3 - 4ja - a^2}{3 - 4ja - a^2} = 0$$

$$RY_{21} + Y_{22}.R.(1 - j6a - 5a^2 + ja^3) + 3 - 4ja - a^2 = 0$$

$$RY_{21} + Y_{22}.R. - 5Y_{22}.R + 3 - a^2 = 0 \text{ and } -j6aY_{22}.R + jY_{22}.Ra^3 - 4ja = 0$$

4) Expression of the oscillation frequency of the circuit

$$ja[-6Y_{22}.R + Y_{22}.Ra^2 - 4] = 0$$

$$a = \frac{1}{RC\omega}; \quad a^2 = \frac{1}{(RC\omega_0)^2} = \frac{6Y_{22}R + 4}{Y_{22}R}$$

$$\omega_0 = \frac{1}{RC}\sqrt{\frac{Y_{22}R}{6Y22R + 4}}$$

When: $Y_{22}R \gg 1;$ $\quad \omega_0 = \frac{1}{RC}\sqrt{\frac{Y_{22}R}{6Y22R + 4}} \cong \frac{1}{\sqrt{6}.RC}$

SOLUTION TO EXERCISE 14

1) Type of oscillator, role of the self-inductance L_1 and role of capacitor C_0

The oscillator under study is a Colpitts oscillator. The self-inductance L_1 is a choke: it allows the protection of supply against rapid variations. The capacitor of capacitance C_0 is a short circuit at the working frequency.

2) Transistor circuit configuration

The transistor is connected in a common base. The capacitor C_0 provides the grounding of the base under dynamic conditions.

3) Dynamic equivalent diagram of the circuit – amplifier chain and feedback loop (see Figure E14.2)

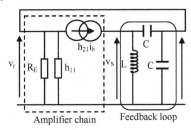

Amplifier chain   Feedback loop

Figure E14.2. *Equivalent diagram of the proposed circuit*

Since resistance R_E is considered very high compared to the input impedance h_{11} of the transistor, its effect is negligible and the equivalent diagram comes down to the circuit shown in Figure E14.3.

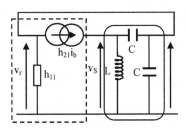

Figure E14.3. *Simplified equivalent diagram*

4) Calculation of the parameters Y_{ij} of the passive quadripole

The passive feedback circuit is represented in Figure E14.4.

$$i_1 = Y_{11}v_1 = Y_{12}v_2$$

$$i_2 = Y_{21}v_1 + Y_{22}v_2$$

$$i_1 = \frac{v_1}{jL\omega} + jC\omega(v_1 - v_2)$$
$$i_2 = jC\omega v_2 + jC\omega(v_2 - v_1)$$

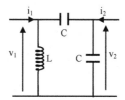

Figure E14.4. *Diagram of the passive feedback circuit*

Based on these equations, the following can be deduced:

$$Y_{11} = jC\omega + \frac{1}{jL\omega}; \ Y_{22} = 2jC\omega$$

$$Y_{12} = Y_{21} = -jC\omega;$$

5) Operating equation of the oscillator and determination of the oscillation frequency

5.1) Operating equation

$$-Y_{21}(h_{21} - h_{12}) + Y_{11} + h_{22} + \Delta h Y_{22} + \Delta Y h_{11} = 0; \ h_{12} \cong 0; \ h_{22} \cong 0; \ \Delta h = 0$$

The operating equation is then reduced to the following relation:

$$-Y_{21}h_{21} + Y_{11} + \Delta Y h_{11} = 0; \quad \text{with: } \Delta Y = Y_{11}Y_{22} - Y_{12}Y_{21}$$

$$\Delta Y = \left[\left[jC\omega + \frac{1}{jL\omega} \right] 2jC\omega \right] + [C\omega]^2 = -2[C\omega]^2 + \frac{2C}{L} + [C\omega]^2 = -[C\omega]^2 + \frac{2C}{L}$$

The operating equation finally becomes:

$$h_{21}C\omega + C\omega - \frac{1}{L\omega} + jh_{11}\left[-\left[C\omega\right]^2 + \frac{2C}{L}\right] = 0$$

5.2) Oscillation frequency F_0

In order to find F_0, it suffices to cancel out the imaginary part of the operating equation:

$$jh_{11}\left[-\left[C\omega_0\right]^2 + \frac{2C}{L}\right] = 0 \Rightarrow \left[-\left[C\omega_0\right]^2 + \frac{2C}{L}\right] = 0$$

$$\left[\omega_0\right]^2 = \frac{2}{LC}; \quad \Rightarrow F_0 = \sqrt{\frac{2}{LC}}$$

SOLUTION TO EXERCISE 15

1) Admittance of quartz crystal

Given the equivalent diagram of the quartz crystal (see Figure E15.2), the following can be noted:

$$Z_1 = R + j\left(L\omega - \frac{1}{C_1\omega}\right); \quad Z_2 = \frac{1}{jC_2\omega};$$

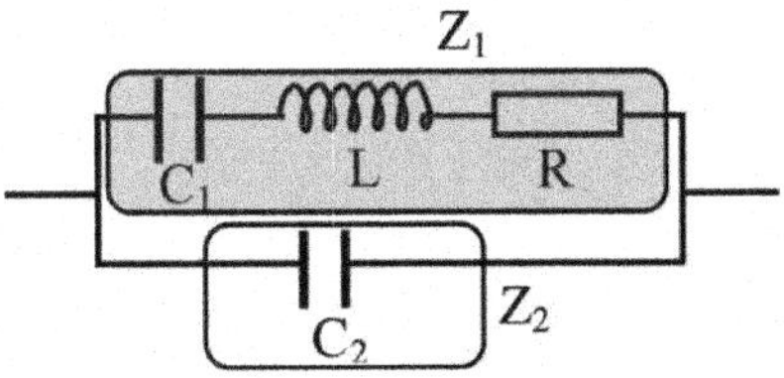

Figure E15.2. *Equivalent diagram of quartz crystal*

The quartz crystal is formed by the parallel connection of two impedances Z_1 and Z_2.

$$Z = \frac{Z_1 Z_2}{Z_1 + Z_2} \Rightarrow Y = \frac{Z_1 + Z_2}{Z_1 Z_2}$$

$$Y = \frac{R + j\left(L\omega - \dfrac{1}{C_1\omega}\right) + \dfrac{1}{jC_2\omega}}{\left(R + j\left(L\omega - \dfrac{1}{C_1\omega}\right)\right)\dfrac{1}{jC_2\omega}} = \frac{jRC_2\omega - C_2\omega\left(L\omega - \dfrac{1}{C_1\omega}\right) + 1}{R + j\left(L\omega - \dfrac{1}{C_1\omega}\right)}$$

2) Condition for the simplification of the expression of Y according to the given model

In order to obtain the desired expression of Y, it is sufficient to neglect the effect of resistance R ($R \cong 0$).

$$Y = \frac{jRC_1C_2\omega^2 + C_2\omega\left(1 - LC_1\omega^2\right) + C_1\omega}{RC_1\omega - j\left(1 - LC_1\omega^2\right)} = \frac{j\left[(C_2\omega + C_1\omega) - LC_2C_1\omega^3\right]}{\left(1 - LC_1\omega^2\right)}$$

The expression of admittance Y can be written in the following form:

$$Y = j\omega(C_1 + C_2)\frac{1 - \dfrac{LC_1C_2\omega^2}{(C_1 + C_2)}}{1 - LC_1\omega^2}$$

3) Expression of f_p as a function of f_s and value of f_p and f_s

3.1) Expression of f_p as a function of f_s

$$Y = j\omega(C_1 + C_2)\frac{1 - \dfrac{LC_1C_2\omega^2}{(C_1 + C_2)}}{1 - LC_1\omega^2}; \quad Y = j2\pi F(C_1 + C_2)\frac{1 - (F/f_p)^2}{1 - (F/f_s)^2}$$

Identification between the previous two relations yields:

$$f_p = \frac{1}{2\pi}\sqrt{\frac{C_1 + C_2}{LC_1C_2}}, \quad f_s = \frac{1}{2\pi}\sqrt{\frac{1}{LC_1}} \quad \Rightarrow \quad f_p = f_s\left(1 + \frac{C_1}{2C_2}\right)$$

3.2) Values of f_p and f_s

It should be recalled that: $C_1 = 1$ fF, $C_2 = 5$ pF and $L = 250$ mH

$$f_p = \frac{1}{2\pi}\sqrt{\frac{C_1 + C_2}{LC_1C_2}} \cong \frac{1}{2\pi}\sqrt{\frac{1}{LC_1}} = fs = \frac{1}{2\pi}\sqrt{\frac{1}{250.10^{-18}}}$$

$$f_p \cong f_s \qquad f_p \cong f_s = 10\text{MHz}$$

4) Expression of reactance, its sign and nature

The quartz crystal operates at a frequency F_0 with $f_s < F_0 < f_p$. Under these conditions, the expression of reactance presented by the quartz crystal will be searched for. Since the two frequencies f_p and f_s are close ($f_s \cong f_p$), the following can be written:

$$1 - (F_0/f_p)^2 \cong \varepsilon \text{ and } 1 - (F_0/f_s)^2 \cong -\varepsilon \text{ with } \varepsilon > 0$$

Consequently, the following can be written:

$$Y = j2\pi F(C_1 + C_2)\frac{1 - (F_0/f_p)^2}{1 - (F_0/f_S)^2} = j2\pi F(C_1 + C_2)\frac{-\varepsilon}{\varepsilon}$$

$$Y = jX = -j\omega(C_1 + C_2) \qquad X = -\omega(C_1 + C_2)$$

Since ω, C_1 and C_2 are positive quantities, the sign of reactance is negative. It can then be deduced that this reactance has a self-inductance nature.

5) Calculation of the quality factor of quartz crystal and comparison to an LC circuit

The quality factor of quartz crystal is given by $Q_{quartz} = (1/2\pi RC_1F_0)$.

$R = 10\ \Omega$, $C_1 = 1$ fF and $F_0 = 10$ MHz: $Q_{quartz} \cong 16.10^5$

$Q_{quartz} \gg Q_{LC}$: The quartz crystal-based oscillator exhibits better frequency stability than an oscillator based on a simple LC circuit.

SOLUTION TO EXERCISE 16

1) Equivalent diagram of this circuit and role of the capacitor C_c

1.1) Equivalent diagram (Figure E16.3)

1.2) Role of the capacitor C_c

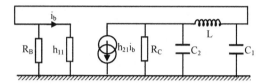

Figure E16.3. *Equivalent diagram of the oscillator*

The capacitor C_c prevents the continuous voltage applied to the collector from being transmitted to the base and thus disturbs the polarization of the transistor.

2) Identification of the elements in Figure E16.3 with the elements of the proposed equivalent diagram (Figure E16.4)

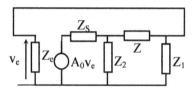

Figure E16.4.

The analysis of the two diagrams (Figures E16.5) allows the identification of various parameters:

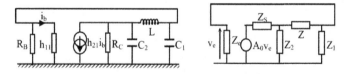

Figure E16.5.

In fact, it can be readily noted that:

$$Z_e = \frac{h_{11}.R_B}{h_{11}+R_B}; \quad Z_S = R_C; \quad Z_1 = \frac{1}{jC_1\omega}$$

$$Z_2 = \frac{1}{jC_2\omega}; \quad Z = jL\omega \quad A_0v_e = -h_{21}R_Ci_b$$

3) Parameters Z_{ij}, operating equation and oscillation frequency

3.1) Parameters Z_{ij} of the passive quadripole

The passive quadripole is represented by the diagram shown in Figure E16.6:

$$v_1 = Z_{11}.i_1 + Z_{12}.i_2$$

$$v_2 = Z_{21}.i_1 + Z_{22}.i_2$$

$$v_1 = Z_2 i_{C2} = Z(i_1 - i_{C2}) + Z_1(i_1 + i_2 - i_{C2})$$

$$v_2 = Z_1. (i_1 + i_2 - i_{C2}) = -Z(i_1 - i_{C2}) + Z_2.i_{C2}$$

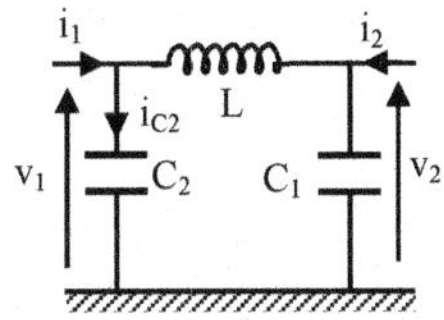

Figure E16.6. *Diagram of the passive feedback quadripole*

The impedances Z_1, Z_2 and Z have already been defined:

$$Z_1 = \frac{1}{jC_1\omega}; \quad Z_2 = \frac{1}{jC_2\omega}; \quad Z = jL\omega$$

$$Z_2.i_{C2} = Z.(i_1 - i_{C2}) + Z_1.(i_1 + i_2 - i_{C2}) \Rightarrow i_{C2}.(Z + Z_1 + Z_2) = (Z + Z_1).i_1 + Z_1.i_2$$

$$i_{C2} = \frac{(Z + Z_1)}{(Z + Z_1 + Z_2)}i_1 + \frac{Z_1}{(Z + Z_1 + Z_2)}i_2$$

Using the expression of i_{C2}, v_1 can be determined as a function of i_1 and i_2:

$$v_1 = Z_2 i_{c2} = \frac{Z_2.(Z + Z_1)}{(Z + Z_1 + Z_2)}i_1 + \frac{Z_1.Z_2}{(Z + Z_1 + Z_2)}i_2$$

The expressions of Z_{11} and Z_{12} can then be deduced. The focus being here on a passive linear quadripole, Z_{21} can also be deduced:

$$Z_{11} = \frac{Z_2.(Z + Z_1)}{(Z + Z_1 + Z_2)}; \quad Z_{12} = Z_{21} = \frac{Z_2.Z_1}{(Z + Z_1 + Z_2)}$$

Using the expression of voltage v_2 as a function of i_1 and i_2 facilitates the determination of Z_{22} and verification that $Z_{12} = Z_{21}$.

$$v_2 = Z_1.(i_1 + i_2 - i_{C2}); \quad v_2 = Z_1 i_1 + Z_1 i_2 - \frac{(Z + Z_1)}{(Z + Z_1 + Z_2)} i_1 + \frac{Z_1}{(Z + Z_1 + Z_2)} i_2$$

$$v_2 = \frac{(Z_1.Z_2)}{(Z + Z_1 + Z_2)} i_1 + \frac{Z_1.(Z + Z_2)}{(Z + Z_1 + Z_2)} i_2$$

Hence: $Z_{21} = \dfrac{(Z_1.Z_2)}{Z + Z_1 + Z_2} = Z_{12}$ and $Z_{22} = \dfrac{Z_1.(Z + Z_2)}{Z + Z_1 + Z_2}$

The following are finally obtained:

$$Z_{11} = \frac{-j(1 - LC_1\omega^2)}{\omega\left[(C_1 + C_2) - LC_1C_2\omega^2\right]}; \quad Z_{22} = \frac{-j(1 - LC_2\omega^2)}{\omega\left[(C_1 + C_2) - LC_1C_2\omega^2\right]}$$

$$Z_{12} = Z_{21} = \frac{-j}{\omega\left[(C_1 + C_2) - LC_1C_2\omega^2\right]}$$

3.2) Operating equation featuring R_B and R_c

$$Z_{21}(h_{21} - h_{12}) + Z_{22} + h'_{11} + \Delta h' Z_{11} + \Delta Z h'_{22} = 0$$

$$h_{12} = 0 \Rightarrow Z_{21}.h_{21} + Z_{22} + h'_{11} + \Delta h'.Z_{11} + \Delta Z.h'_{22} = 0$$

The following should be noted in this equation:

$$h'_{11} = \frac{h_{11}.R_B}{h_{11} + R_B}; \quad h'_{22} = \frac{1}{R_C} \text{ and } \Delta h'_{11} = h'_{11}.h'_{22}$$

Finally, taking into account the bias resistances, the operating equation of the oscillator can be written as follows:

$$Z_{21}.h_{21} + Z_{22} + \frac{h_{11}.R_B}{h_{11} + R_B} + (\frac{h_{11}.R_B}{R_C(h_{11} + R_B)} h_{11}.Z_{11}) + \frac{\Delta Z}{R_C} = 0$$

3.3) Oscillation frequency and condition for sustained oscillation

The influence of R_B and R_C is considered negligible. Since $h_{12} = 0$ and $h_{22} = 0$, $\Delta h = 0$.

Under these conditions, the operating equation of the oscillator becomes:

$$Z_{21}(h_{21} - h_{12}) + Z_{22} + h_{11} + \Delta h Z_{11} + \Delta Z h_{22} = h_{21}.Z_{21} + Z_{22} + h_{11} = 0$$

$$h_{21}.Z_{21} + Z_{22} + h_{11} = 0$$

When Z_{21} and Z_{22} are replaced by their equivalent expression, the following relations are obtained:

$$Z_{22} = \frac{-j(1-LC_2\omega^2)}{\omega\left[(C_1+C_2)-LC_1C_2\omega^2\right]}; \quad Z_{12} = Z_{21} = \frac{-j}{\omega\left[(C_1+C_2)-LC_1C_2\omega^2\right]}$$

$$h_{21}\frac{-j}{\omega\left[(C_1+C_2)-LC_1C_2\omega^2\right]} + \frac{-j(1-LC_2\omega^2)}{\omega\left[(C_1+C_2)-LC_1C_2\omega^2\right]} + h_{11} = 0$$

$$h_{21} + (1-LC_2\omega^2) + jh_{11}\omega\left[(C_1+C_2)-LC_1C_2\omega^2\right] = 0$$

In order to obtain the oscillation frequency, it is sufficient to cancel out the imaginary part of the operating equation:

$$h_{21} + (1-LC_2\omega^2) + jh_{11}\omega\left[(C_1+C_2)-LC_1C_2\omega^2\right] = 0$$

$$\left[(C_1+C_2)-LC_1C_2\omega_0^2\right] = 0$$

$$F_0 = \frac{\omega_0}{2\pi} = \frac{1}{2\pi}\sqrt{\frac{C_1+C_2}{LC_1C_2}} = \frac{1}{2\pi}\frac{1}{\sqrt{LC_{eq}}} \quad \text{with } C_{eq} = \frac{C_1.C_2}{C_1+C_2}$$

SOLUTION TO EXERCISE 17

1) Calculation of the impedance of the quartz crystal

$$Z = \frac{1+jC_s\omega(r+jL\omega)}{j(C_s+C_p)\omega - C_sC_p\omega^2(r+jL\omega)}$$

2) Simplified expression of Z (r = 0) and value under steady-state condition

$$Z = \frac{1 - C_s L\omega^2}{j(C_s + C_p)\omega - jC_s C_p L\omega^3} \; ; \quad Z_{\omega=0} \to \infty$$

3) Simplified and developed expression

$$Z = \frac{1 - C_s L\omega^2}{j(C_s + C_p)\omega - jC_s C_p L\omega^3}$$

$$Z = \frac{1}{j(C_s + C_p)\omega} \frac{1 - \left[\dfrac{\omega}{\omega_1}\right]^2}{1 - \dfrac{C_s C_p L\omega^2}{C_s + C_p}} = \frac{1}{jC_{eq}\omega} \frac{1 - \left[\dfrac{\omega}{\omega_1}\right]^2}{1 - \left[\dfrac{\omega}{\omega_2}\right]^2}$$

4) Expressions of C_{eq}, ω_1 and ω_2. Let us prove that $\omega_1 < \omega_2$. Significance of ω_1 and ω_2.

$$C_{eq} = (C_s + C_p); \qquad \omega_1 = \sqrt{\frac{1}{LC_S}} \qquad \omega_2 = \sqrt{\frac{(C_s + C_p)}{LC_s C_p}}$$

$C_p \gg C_s$

$$\omega_1 = \sqrt{\frac{1}{LC_S}} \qquad \omega_2 = \sqrt{\frac{(C_s + C_p)}{LC_s C_p}}$$

The two angular frequencies can then be expressed as functions of each other:

$$\omega_2 = \omega_1 \sqrt{\frac{C_s + C_p}{C_p}} = \omega_1 \left[1 + \frac{C_s}{C_p}\right]^{1/2} \cong \omega_1\left[1 + \frac{C_s}{2C_p}\right], \qquad \omega_2 > \omega_1$$

where ω_1 is the resonance angular frequency of the LC_s circuit (series circuit) and ω_2 is the resonance angular frequency of the L-C_P-C_s circuit (parallel circuit).

5) Variations of the module and argument of Z as a function of ω. Natures of the impedance of the quartz crystal in the various ranges separated by ω_1 and ω_2.

5.1) Module of the impedance of quartz crystal

$$Z = \frac{1}{jC_{eq}\omega} \frac{\left[1 - \left[\dfrac{\omega}{\omega_1}\right]^2\right]}{\left[1 - \left[\dfrac{\omega}{\omega_2}\right]^2\right]} \quad \Rightarrow \quad |Z| = \frac{1}{C_{eq}\omega} \frac{\left[1 - \left[\dfrac{\omega}{\omega_1}\right]^2\right]}{\left[1 - \left[\dfrac{\omega}{\omega_2}\right]^2\right]}$$

b) Argument of the impedance of quartz crystal

$$Arg(Z) = \begin{cases} \phi = -\pi/2 & \omega < \omega_1 \\ \phi = \pi/2 & \omega_1 < \omega < \omega_2 \\ \phi = -\pi/2 & \omega > \omega_2 \\ \phi = 0 & \omega = \omega_1 \quad et \quad \omega = \omega_2 \end{cases}$$

c) Representation of the module and phase of Z (Figure E17.2) and nature of the impedance Z

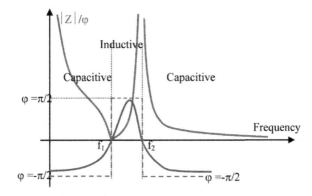

Figure E17.2. *Evolution of the impedance of quartz crystal in module and phase. For a color version of this figure, see www.iste.co.uk/haraoubia/nonlinear1.zip*

5

Circuits in Switching Mode

5.1. Basic elements

The reaction of circuits to square or rectangular pulsed signals is a major preoccupation. Therefore, it is worth recalling several fundamental notions related to this type of signal.

A square or a rectangular signal (Figure 5.1) is defined by several parameters, the most important of which are as follows:

– high state and low state duration;

– cyclic ratio;

– average value;

– frequency, rising edge, falling edge, etc.

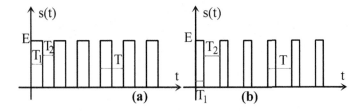

Figure 5.1. *a) Square signal and b) rectangular signal*

High state and low state durations for each type of signal (square or rectangular) are defined by quantities T_1 and T_2, respectively. The relation $T_1 = T_2$ is always valid for the square signal.

The cyclic ratio (r) is defined by: $r = \dfrac{T_1}{T_1 + T_2}$

For the square signal, the value of the cyclic ratio is equal to 50%.

In the general case of an arbitrary signal v(t), whose period is T, the average value (V_{av}) is defined by the following relation:

$$V_{av} = \frac{1}{T}\int_t^{t+T} v(t)dt$$

For the signals represented in Figure 5.1, period T and the average value (V_{av}) are expressed by the following relations:

$$T = T_1 + T_2 \qquad \text{and} \qquad V_{av} = E\frac{T_1}{T}$$

For a unidirectional square signal, the average value is half the maximal value of the signal in question (case of the signal in Figure 5.1(a)): $V_{av} = (E/2)$.

5.2. Behavior of a capacitor in a circuit

The simplest circuit that comprises one capacitor is represented by the diagram in Figure 5.2. The objective is to find how the voltage across capacitor "C" evolves in time.

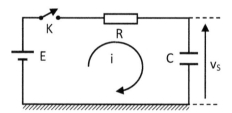

Figure 5.2. *RC circuit and charge of capacitor C*

When switch K is closed at instant t = 0, capacitor C charges through resistance R. Hence, the following can be written as follows:

$$E = Ri + \frac{1}{C}\int_0^t idt\,; \quad v_s = \frac{1}{C}\int_0^t idt$$

Replacing the current "i" with its equivalent expression as a function of output voltage v_s yields the following expression:

$$E = RC\frac{dv_s}{dt} + v_s$$

A particular solution is $v_s = E$. The solution to the differential equation without the second member:

$$RC\frac{dv_s}{dt} + v_s = 0$$

yields

$$v_s = A.e^{-\frac{t}{RC}}$$

"A" is a constant that can be determined according to the initial conditions. The general solution is:

$$v_s = A.e^{-\frac{t}{RC}} + E$$

at $t = 0$; $v_s = 0 \Rightarrow A = -E$

$$v_s = E(1 - e^{-\frac{t}{RC}})$$

The evolution of voltage across the capacitor is shown in Figure 5.3.

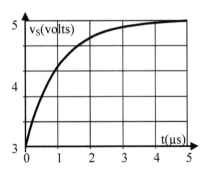

Figure 5.3. Charge of capacitor C through resistance R and evolution of voltage across C. E = 5V, R = 100Ω and C = 10 nF, τ = R.C = 1 μs

The capacitor is charged at voltage E, equal to the voltage supplied by the generator. When the capacitor is fully charged, the current (i) across the mesh is zero.

It should be noted that the capacitor is fully charged after a duration that is nearly five times its time constant $\tau = RC$.

This parameter is a very important data in the circuits that have capacitors and resistances in their structure.

From now on, it can therefore be stated that a capacitor charges rapidly when its time constant is small.

When the capacitor is first charged at a potential equal to E and is connected in the circuit as shown in Figure 5.4 (the switch is closed at an instant chosen equal to zero), the capacitor discharges across resistance R according to the law:

$$V_s = E.e^{-\frac{t}{RC}}$$

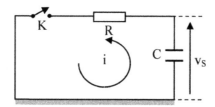

Figure 5.4. *Circuit for the discharge of capacitor C*

The variations of discharge voltage are schematically represented in Figure 5.5.

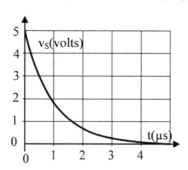

Figure 5.5. *Evolution of discharge voltage across capacitor C*

5.3. RC circuits in switching mode

5.3.1. *Case of a low-pass cell – integrating circuit*

A square signal (Figure 5.6(a) is applied at the input of a low-pass RC circuit, Figure 5.6(b)). The task at hand is to determine the nature and the shape of the signal at the output of this RC circuit (voltage across the capacitor). As already noted, the charge and discharge voltages of capacitor C evolve according to the following relations:

$$V_{ch\,arg\,e} = V_{ch} = E(1 - e^{\frac{-t}{RC}})$$

$$V_{disch\,arg\,e} = V_{dec} = E.e^{\frac{-t}{RC}}$$

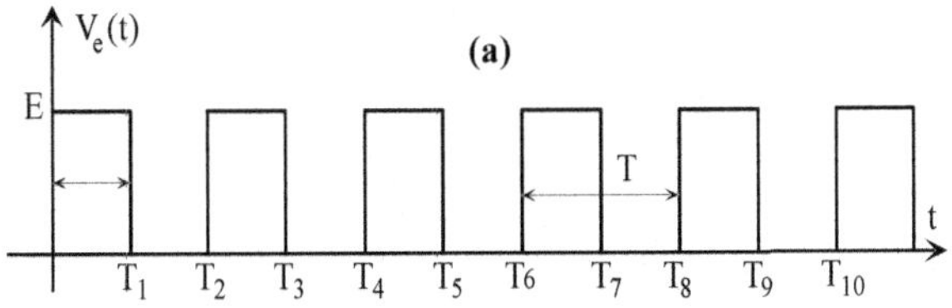

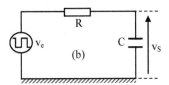

Figure 5.6(a). *Square signal applied at the input of an RC circuit*

Figure 5.6(b). *A low-pass RC circuit under study*

These relations are only valid for the fixed initial conditions, as follows:

– for the charge, the capacitor is considered to be fully discharged at the initial instant;

– for the discharge, the capacitor is considered to be charged at potential +E at the start;

– when a unidirectional square signal of amplitude E (Figure 5.6(a)) is applied to a low-pass RC circuit (Figure 5.6(b)), it is important to rewrite the equations for the capacitor's charge and discharge, taking into account the possible initial voltages that may exist;

– the shape of the output signal depends on the time constant (τ = R.C) of the circuit in question and on the initial voltages, taken during successive charges and discharges of capacitor C. A capacitor charges according to a law of which the general expression is:

$$v_s = V_{ch} = Ae^{\frac{-t}{RC}} + B;$$

– "A" and "B" are two constants that depend on the boundary conditions of the charge. Thus, at instant t = 0 (for the charge), an initial voltage denoted V_{ic} is assumed to exist;

– $A + B = V_{ic}$;

– when the time "t" tends to infinity, the charge of the capacitor tends to the maximal value of the voltage applied;

– in the case that is of interest here, this value is equal to E, which is the amplitude of the square signal;

– $B = E$, $A = (V_{ic} - E)$;

– finally, the general expression related to the charge of a capacitor submitted to a square signal of amplitude E is defined by:

$$v_s = V_{ch} = E - (E - V_{ic})e^{\frac{-t}{RC}};$$

– the same type of relation is applicable to the discharge of the capacitor:

$$v_s = V_{dec} = De^{\frac{-t}{RC}} + F;$$

– quantities "D" and "F" depend on the boundary conditions related to capacitor discharge;

– thus, for an infinite time, the capacitor tends to fully discharge and reach the lowest value applied to it. In this case, this value is zero: F = 0;

– on the other hand, for the instant when discharge starts, the initial voltage related to discharge should be taken into account (this voltage is in fact the final voltage of the charge that preceded this discharge): $D = V_{id}$;

– This leads to the relation for discharge voltage in the general case:

$$v_s = V_{dec} = V_{id}e^{\frac{-t}{RC}};$$

– to accurately follow the evolution of voltage v_s, let us set the numerical values of the quantities involved;

– for example, the following values have been chosen: E = 5 volts; R = 1 KΩ; C = 1 μF; τ = RC = 1 ms) and the frequency "f" of the square signal is set at 500 Hz (T = 2 ms).

– Table 5.1 indicates the variations of output voltage in time:

Capacitor state	Time evolution	Evolution of the values of voltage v_s (in volts)
Initial	0	0
Charge	$T_1 = 1$ ms	3.16
Discharge	$T_2 = 2$ ms	1.16
Charge	$T_3 = 3$ ms	3.58
Discharge	$T_4 = 4$ ms	1.32
Charge	$T_5 = 5$ ms	3.64
Discharge	$T_6 = 6$ ms	1.34
Charge	$T_7 = 7$ ms	3.65
Discharge	$T_8 = 8$ ms	1.344
Charge	$T_9 = 9$ ms	3.655
Discharge	$T_{10} = 10$ ms	1.345

Table 5.1. *Variations of the output voltage (across the capacitor) in time for RC = T/2*

– the variations of the output signal in relation with the input signal are shown in Figure 5.7:

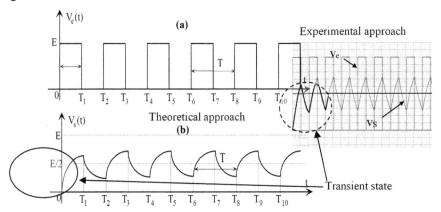

Figure 5.7. *(b) Evolution of the output voltage in relation with (a) input voltage when: E = 5 Volts; R = 1KΩ; C = 1μF and f = 500 Hz*

– it can be noted that after a limited time (related to the transient state), the capacitor charge and the discharge voltages evolve around the average value (E/2) of the input square signal;

– for a better study of this phenomenon, let us consider two values of time constants τ_1 and τ_2 which are very distant from one another and which also meet certain conditions with respect to the period of the input signal;

– $\tau_1 \ll T$ and $\tau_2 \gg T$

– T is the period of the input signal.

5.3.1.1. *Case where the time constant is very low compared to the period*

– $\tau_1 \ll T$ ($\tau_1 = RC = 0.10$ ms).

– The following are given: E = 5 Volts; R = 1000 Ω; C = 0.1 μF and T = 2 ms.

– The values taken by the output voltage in time under these conditions are summarized in Table 5.2.

Capacitor state	Time evolution	Evolution of voltage values v_s (in volts)
Initial	0	0
Charge	$T_1 = 1$ ms	5
Discharge	$T_2 = 2$ ms	0
Charge	$T_3 = 3$ ms	5
Discharge	$T_4 = 4$ ms	0
Charge	$T_5 = 5$ ms	5
Discharge	$T_6 = 6$ ms	0

Table 5.2. *Variations of the output voltage (across the capacitor) in time for RC << T*

– It can be noted that the output voltage evolves around the average value of the input signal. This is also the average value of the output signal. Furthermore, the extreme values of the output signal (see the diagram in Figure 5.8) are similar to those of the input signal. Given the low value of the charge and discharge time constant, the capacitor has the time to fully charge and discharge.

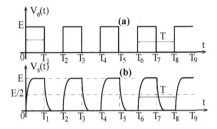

Figure 5.8. *Evolution of the input voltage (a) in relation with the signal across capacitor (b), when the time constant is small ($\tau_1 = RC = 0.10$ ms); $T = 2$ ms*

5.3.1.2. *Case where the time constant is very large compared to the period*

$- \tau_2 \gg T$ ($\tau_2 = RC = 4.7$ ms).

– In this case, the following values are chosen: E = 5 V; R = 4.7 KΩ; C = 1 μF and T = 2 ms. The values taken by the output voltage are summarized in Table 5.3.

Capacitor state	Time evolution	Evolution of the values of voltage (in volts)
Initial	0	0
Charge	$T_1 = 1$ ms	0.96
Discharge	$T_2 = 2$ ms	0.77
Charge	$T_3 = 3$ ms	1.59
Discharge	$T_4 = 4$ ms	1.28
Charge	$T_5 = 5$ ms	1.99
Discharge	$T_6 = 6$ ms	1.6
Charge	$T_7 = 7$ ms	2.26
Discharge	$T_8 = 8$ ms	1.82
Charge	$T_9 = 9$ ms	2.43
Discharge	$T_{10} = 10$ ms	1.96
Charge	$T_{11} = 11$ ms	2.54
Discharge	$T_{12} = 12$ ms	2.05
Charge	$T_{13} = 13$ ms	2.61
Discharge	$T_{14} = 14$ ms	2.11

Table 5.3. *Variations of the output voltage (across the capacitor) in time for RC $\gg$ T*

– The evolution of the output signal in time is represented by the diagram in Figure 5.9. In this case, it can be noted that reaching a steady state requires time. The voltage across the capacitor tends to stabilize around a value that is very close to the average value of the input signal. If the time constant of the RC circuit is further increased and for a steady state that has been reached, the curve of the obtained signal is piecewise linear (charge and discharge of the capacitor).

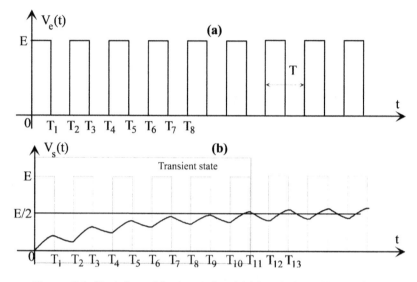

Figure 5.9. *Evolution of the input signal (a) in relation with the signal across the capacitor (b) if $\tau_2 = RC = 4.7$ ms and $T = 2$ ms*

Indeed, an integration of the input signal has occurred. Let us therefore write the relation between the input voltage and the resulting output voltage:

– Case of the charge: $v_e = E$

$$V_s = V_{ch} = E - (E - V_{ic})e^{\frac{-t}{RC}}$$

When the time constant $\tau_2 = R.C$ is very large compared to the signal period and consequently to the duration of the high state, the exponential term can be expanded. A first order expansion yields:

$$v_s = V_{ch} = E - (E - V_{ic})(1 - \frac{t}{RC}) = V_{ic} + (E - V_{ic})\frac{t}{RC}$$

It should be recalled that V_{ic} is the initial voltage across the capacitor when the latter starts to charge.

– Case of discharge: $v_e = 0$

$$v_s = V_{dec} = V_{id} e^{\frac{-t}{RC}}$$

An expansion of the exponential term, similar to the above, leads to the following:

$$v_s = V_{dec} = V_{id}(1 - \frac{t}{RC})$$

It can be noted that the output voltage is a linear function of time. An integration of the input signal has occurred. When the time constant of the RC circuit is large, the circuit realizes the integration function.

A representation of the output signal in a steady state in relation with the input signal applied to an integrating circuit is shown in Figure 5.10.

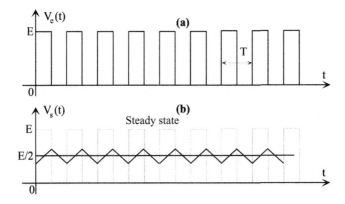

Figure 5.10. *The input signal (a) and the output signal (b) of an integrating circuit in a steady state*

5.3.2. *Case of a high-pass cell – differentiating circuit*

The RC circuit to be studied in this context is a high-pass cell. It is schematically represented in Figure 5.11. In this case, the output is taken across resistance R. The input signal is a square signal. The shape of the output signal depends on the constant $\tau = R.C$ of the circuit in question.

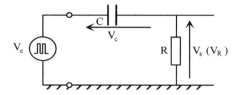

Figure 5.11. *Differentiating RC circuit*

Three cases can be considered with respect to the time constant and to period T of the input signal, as follows:

$\tau_1 = RC \ll T$;

$\tau_2 = RC \cong T$;

$\tau_3 = RC \gg T$.

5.3.2.1. *Case of a very small time constant*

$\tau = \tau_1 \ll T$:

Under these conditions, the capacitor has the time to fully charge and discharge. At the arrival of the first rising edge (zero transition time), the capacitor instantaneously transmits this variation across the resistance and then starts to charge.

Kirchhoff's second law for the RC circuit under study yields:

$v_e = v_c + v_R$ and $v_R = v_e - v_c$

v_c and v_R are voltages across the capacitor and the resistance, respectively. Naturally, the voltage across the resistance is the output voltage:

$$v_c + v_R = \begin{cases} E \text{ for } 0<t<(T/2) \\ E \text{ for } (T/2)<t<T \end{cases}$$

The expression of the evolution of voltage across the capacitor has been previously defined:

$$v_c = \begin{cases} E-(E-V_{ic})e^{\frac{-t}{RC}} \text{ for } 0<t<(T/2) \\ V_{id}.e^{\frac{-t}{RC}} \quad \text{ for } (T/2)<t<T \end{cases}$$

Let us recall that V_{ic} and V_{id} are the initial capacitor charge and discharge voltages, respectively.

These two voltages depend on whether the capacitor is able or not to charge rapidly.

The expression of the output voltage (voltage across resistance R) is defined as follows:

$$V_R = \begin{cases} (E - V_{ic})e^{\frac{-t}{RC}} & \text{for} \quad 0 < t < (T/2) \\ -V_{id}.e^{\frac{-t}{RC}} & \text{for} \quad (T/2) < t < T \end{cases}$$

If the capacitor has the time to fully charge and discharge (the time constant is very small compared to the value of the period of the input signal), it can be stated that:

$$V_{ic} = 0; \; V_{id} = E;$$

$$V_R = \begin{cases} Ee^{\frac{-t}{RC}} & \text{for} \quad 0 < t < (T/2) \\ -Ee^{\frac{-t}{RC}}) & \text{for} \quad (T/2) < t < T \end{cases}$$

The representation of the output signal in relation to the input signal in the case when the time constant is small compared to period T of the input signal, is shown in Figure 5.12. There has been a derivation of the input signal.

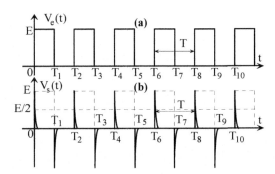

Figure 5.12. *A representation of the input signal a) and the output signal b) when the time constant $\tau_1 = RC \ll T$*

5.3.2.2. *Case of a very large time constant*

$$\tau = \tau_2 \gg (T/2):$$

The capacitor does not have enough time to either charge or discharge. Consequently, the capacitor will partially charge. The output voltage depends on this partial charge.

The output voltage is defined by the following relations:

$$v_s = v_R = \begin{cases} (E - V_{ic}).e^{\frac{-t}{RC}} & \text{for} \quad 0<t<(T/2) \\ -V_{id}.e^{\frac{-t}{RC}} & \text{for} \quad (T/2)<t<T \end{cases}$$

In view of an overall representation of the evolution of the output voltage v_s, the various values that this voltage takes in time are calculated (see Table 5.4). For this purpose, the following values are set: $E = 5$ V; $(T/2) = 1$ ms; $\tau_1 = 4.7$ ms.

Initial value of the voltage involved	Time evolution	Evolution of v_c (in volts)	Evolution of voltage v_s in volts
Initial	0	0	E = 5
Vic = 0	$T_1 = 1$ ms	0.96	4.04
Vid = 0.96	$T_2 = 2$ ms	0.77	−0.77
Vic = 0.77	$T_3 = 3$ ms	1.59	3.41
Vid = 1.59	$T_4 = 4$ ms	1.28	−1.28
Vic = 1.28	$T_5 = 5$ ms	1.99	3.01
Vid = 1.99	$T_6 = 6$ ms	1.6	−1.6
Vic = 1.60	$T_7 = 7$ ms	2.26	2.74
Vid = 2.26	$T_8 = 8$ ms	1.82	−1.82
Vic = 1.82	$T_9 = 9$ ms	2.43	2.57
Vid = 2.43	$T_{10} = 10$ ms	1.96	−1.96
Vic = 1.96	$T_{11} = 11$ ms	2.54	2.46
Vid = 2.54	$T_{12} = 12$ ms	2.05	−2.05
Vic = 2.05	$T_{13} = 13$ ms	2.61	2.39
Vid = 2.61	$T_{14} = 14$ ms	2.11	−2.11
Vic = 2.11	$T_{15} = 15$ ms	2.66	2.34
Vid = 2.66	$T_{16} = 16$ ms	2.15	−2.15
Vic = 2.15	$T_{17} = 17$ ms	2.70	2.30
Vid = 2.70	$T_{18} = 18$ ms	2.18	−2.18
Vic = 2.18	$T_{19} = 19$ ms	2.72	2.28
Vid = 2.72	$T_{20} = 20$ ms	2.20	−2.20
Vic = 2.20	$T_{21} = 21$ ms	2.73	2.27

Table 5.4. *Variations of the output voltage (across the resistance) in time for RC >> T/2*

It can be noted that the following relation is verified at any moment:

$$v_c + v_R = v_e$$

This is quite normal, since it is the law that governs the circuit under study. It should be recalled that

$$v_e = \begin{cases} E & \text{for} \quad 0<t<(T/2) \\ 0 & \text{for} \quad (T/2)<t<T \end{cases}$$

The curves of the input and output signals are represented in Figures 5.13(a) and (b), respectively.

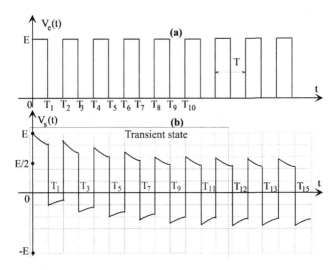

Figure 5.13. *a) Variation of the input signal and b) variation of the output signal of a high-pass RC circuit when the time constant RC >> T*

It can be noted that the continuous component of the input signal has not been transmitted toward the output, being blocked by the capacitor. The shape of the output signal closely resembles the shape of a square signal. Its average value is zero (it is perfectly square if the value of the time constant is further increased compared to the period of the input signal). It can therefore be stated that the capacitor transmits the rapid variations of a signal instantaneously and blocks any continuous component. This circuit is generally used in coupling functions. For this purpose, the time constant has been chosen to be sufficiently large in order to avoid the possible amplitude distortions of the output signal. An example of the output signal when the time constant is very large compared to the period (for example, RC = 20T) is shown schematically in Figure 5.14. This represents the curve of the output signal under an established steady state. The transient state has been omitted.

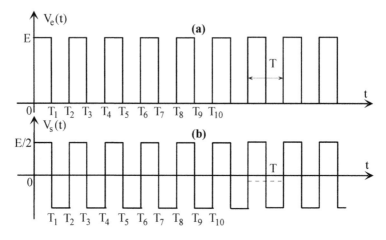

Figure 5.14. *a) The input signal and b) the output signal under a steady state (the time constant of the RC circuit is very large compared to the period of the input signal)*

IN SUMMARY.−

1) An integrating RC circuit favors the transfer of the continuous component with respect to a high frequency signal. An RC circuit is qualified as an integrator when its time constant is large compared to the period of the input signal. The output is considered across the capacitor.

2) A differentiating RC circuit does not allow the passage of the continuous component from the input to the output. The output of a differentiator is considered across the resistance. The RC circuit cannot be a differentiator circuit, unless its time constant is very small compared to the period of the input signal.

3) A capacitor instantaneously transmits abrupt variations of a signal. It opposes the passage of a continuous current.

5.4. Bipolar transistor in switching mode

5.4.1. *Bipolar transistor characteristics*

The transistor used in switching mode must uniquely operate either in a blocked state or in a saturated state. The transistor used for this function is generally connected as a common emitter (Figure 5.15).

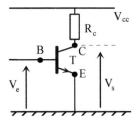

Figure 5.15. *A bipolar transistor connected as a common emitter*

The static load line is defined by the following relation:

$$I_C = \frac{V_{cc} - V_{CE}}{R_c}$$

The only states of the transistor that operates in switching mode are:

$I_C = 0$ and $V_{CE} = V_{cc}$ (blocked transistor)

$I_C = \dfrac{V_{cc}}{R_c} = I_{CMAX}$ and $V_{CE} \cong 0$ (saturated transistor)

These two points are in fact the two extreme points of the load line (Figure 5.16).

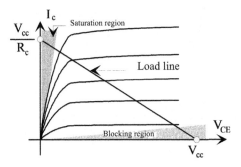

Figure 5.16. *Load line and operating points of a transistor in a switching mode*

The blocked state of the transistor can be obtained when a base-emitter voltage (V_{BE}) below the threshold voltage is applied. For an NPN bipolar transistor, the threshold voltage is around 0.6 volts.

This is the minimal voltage required for the forward bias of the base-emitter junction of the transistor.

Therefore, for an NPN bipolar transistor, the following can be written with approximation:

$V_{BE} < 0.6$ volt; $I_B = 0$; $I_C = 0$ and $V_{CE} = V_{cc}$.

The saturated state is reached when the base-emitter voltage is greater than or equal to the threshold voltage. For the NPN transistor:

$$V_{BE} \geq 0.6 \text{ V}; I_B \neq 0; \ I_C = I_{CMAX} \cong \frac{V_{cc}}{R_c} \text{ and } V_{CE} \cong 0$$

5.4.2. Operation in switching mode

The passage from a blocked state to a saturated state (or vice versa) takes place abruptly and practically instantaneously. When a sinusoidal signal is applied to the base of a transistor that is connected as a common emitter (Figure 5.17), the output signal has a square form.

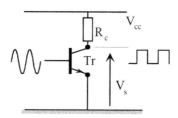

Figure 5.17. *A transistor in a switching mode with a sinusoidal input signal*

Each time the input voltage is above the threshold, a transition from one state to the other occurs at the output, as shown in Figure 5.18.

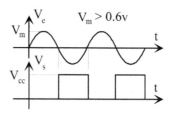

Figure 5.18. *Shape of the output signal in relation to the input signal*

The curve of the output signal cannot be perfectly square, and the rising and falling edges are not as steep as suggested by the diagram shown in Figure 5.18.

The curve of the output signal is rather that defined in Figure 5.19(b). The rising and falling times and the various characteristic times are not zero. The values of these times depend on the technology employed.

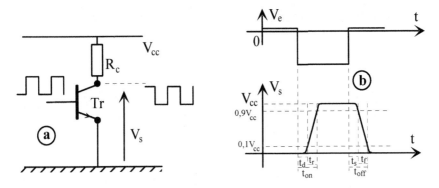

Figure 5.19. a) Switching circuit and b) real curve
of the output signal and characteristic times

The various characteristic times are defined as follows:

t_d, delay time; t_r, rise time; t_{on}, setting time; t_s, storage time; t_f, fall time and t_{off}, lagging time.

5.4.3. Logical functions with a switching transistor

Several logical operations can be realized with bipolar transistors operating in switching mode. As a reference, let us provide several examples of these functions. Among the simplest operations to be realized, the following are worth mentioning: inverter gate, NOT AND or NAND gate, etc.

5.4.3.1. *Inverter gate*

The representative diagram of the inverter gate and the signals involved is shown in Figure 5.20.

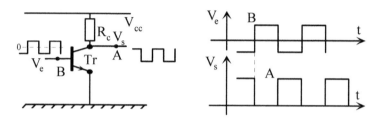

Figure 5.20. *Inverter gate with a bipolar transistor and the signals involved*

When the input is in a low state, the output is in a high state, and vice versa:

$$A = \overline{B}$$

5.4.3.2. *NOT AND or NAND gate*

The electric diagram of this gate using bipolar transistors is shown in Figure 5.21.

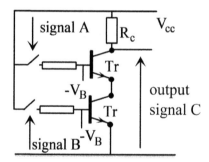

Figure 5.21. *NOT AND or "NAND" gate with a bipolar transistor*

This gate uses two bipolar transistors. With another type of technology, and for reasons of consumption, field effect transistors can be employed.

When choosing one gate and its technology, there is always a compromise to be made between gate rapidity and energy consumption.

Hence: $C = \overline{A.B}$

6

Astable Multivibrators

6.1. Introduction

Similar to any unstable system, an astable circuit requires no input voltage to generate a generally rectangular or square wave, as shown in Figure 6.1. The astable multivibrator uses a positive feedback in order to introduce the instability phenomenon.

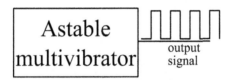

Figure 6.1. *Principle of astable multivibrator*

The Fourier series expansion of the output signal evidences its spectral richness. This can be explained by the fact that the condition for sustained oscillation, also called the "Barkhausen" condition, is met for a wide range of frequencies, hence the name multivibrator.

6.2. Astable multivibrator with transistors

6.2.1. *Introduction*

It is worth noting that the study of a multivibrator realized with bipolar transistors (discrete components) is of very high didactic interest. On the contrary, the practical realization of astable multivibrators with transistors is out-of-date. There are integrated circuits such as operational amplifiers or logic circuits that allow the design of astable circuits that are both very simple and high performing. Furthermore, the development of electronics has allowed the marketing of specialized astable circuits. The operation of these circuits requires only few external passive components (in general, few resistances and capacitors). Therefore, the astable multivibrator with transistors is highly recommended for teaching purposes and for first studies and realizations. The objective is the proper understanding of various phenomena involved in signal generation by this device, which presents a very strong didactic interest. An example of an astable multivibrator with transistors is schematically shown in Figure 6.2.

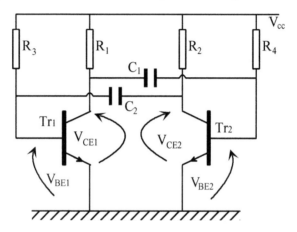

Figure 6.2. *Astable multivibrator with transistors*

6.2.2. *Principle*

Two capacitors C_1 and C_2 are used for the coupling between the base and the collector of two transistors Tr1 and Tr2.

The operation of this circuit relies on the following principle: if one of the two transistors, for example, Tr1, is blocked, then transistor Tr2 is saturated (or vice versa). In fact, when there is a decrease in the collector current I_{C1} of transistor Tr1, there is an increase in collector–emitter voltage.

I_{C1} (decreases): $V_{CE1} = V_{cc} - R_1 I_{C1}$ (increases)

Any abrupt variation of V_{CE1} is instantaneously transmitted through C_1 to the base of Tr2.

V_{CE1} increases; V_{BE2} increases; I_{B2} and I_{C2} increase and then V_{CE2} decreases.

The increase in collector–emitter voltage V_{CE1} of Tr1 leads to the decrease in the collector–emitter voltage of Tr2, as described above.

This cumulative phenomenon goes on until $V_{CE1} \cong V_{cc}$ and $V_{CE2} \cong 0$

Under these conditions, transistor Tr1 is blocked and transistor Tr2 is saturated. The system can by design maintain this state during a certain time and then undergo a switch to reach $V_{CE1} \cong 0$ and $V_{CE2} \cong V_{cc}$

Saturated transistor Tr1 ($V_{CE1} \cong 0$)   Blocked transistor Tr2 ($V_{CE2} \cong V_{cc}$)

The output can be on the collector of either Tr1 or Tr2. The signal obtained at one of these outputs is rectangular or square. The amplitude of the high level is $+V_{cc}$ and that of the low level is around 0 V (in reality, several hundreds of mV). The principle for obtaining a square or rectangular signal is examined in detail in the following section.

6.2.3. *Operating condition*

For the device to really operate as an astable multivibrator, the values of resistances must be appropriately chosen.

For example, when transistor Tr1 is saturated, $V_{CE1} = 0$ and $V_{cc} = R_1 I_{C1}$.

The base current I_{B1} is defined as $I_{B1} = \dfrac{I_{C1}}{\beta}$, where β is the current gain of the transistor Tr1.

The base current of transistor Tr1 at saturation is expressed as follows:

$$I_{B1} = \frac{V_{cc} - V_{BE1}}{R_3} = \frac{I_{C1}}{\beta} = \frac{V_{cc}}{\beta R_1}$$

Hence:

$$\beta \frac{R_1}{R_3} = \frac{V_{cc}}{V_{cc} - V_{BE1}} > 1$$

It can be readily noted that transistor Tr1 cannot reach saturation unless $R_3 < \beta R_1$.

If the value of R_3 is excessively high ($R_3 > \beta R_1$), the maximum base current of Tr1 will be low, similar to the collector current required for bringing transistor Tr1 into the saturation region.

Obviously, the same reasoning applies for transistor Tr2, and the overall operating condition of the astable system is given by:

$$R_3 < \beta R_1 \text{ and } R_4 < \beta R_2$$

6.2.4. Operation

For a more simple explanation of the operation of the astable device with transistors, let us assume that Tr1 is blocked and Tr2 is saturated, but that before reaching saturation, it was in a blocked state.

Tr2 blocked $\rightarrow$ Tr2 saturated

$V_{CE2} \cong V_{cc}$ $\rightarrow$ $V_{CE2} \cong 0$ V

This sudden shift of V_{CE2} from $+V_{cc}$ to 0 V is instantaneously transmitted to the base of Tr1 through the capacitance C_2. This is reflected by a voltage drop across the base of transistor Tr1.

Tr1 saturated $\rightarrow$ Tr1 blocked

$V_{CE1} \cong 0$ $\rightarrow$ $V_{CE1} \cong V_{cc}$

The instant when transistors shift from one state to the other is chosen as the time origin ($t_0 = 0$). In the following, we analyze the shape of signals at various points of the astable device.

When $t = t_0 = 0$:

Tr2 blocked $\rightarrow$ Tr2 saturated

$V_{BE2} \leq 0$ $\rightarrow$ $V_{BE2} \cong 0.6$ V

Transistor Tr1 is blocked and Tr2 is saturated. The right plate of capacitor C_2 is at ground potential.

Therefore, this capacitor charges through resistance R_3 (from t_0), trying to reach the supply voltage V_{cc} applied via R_3 (see Figure 6.3).

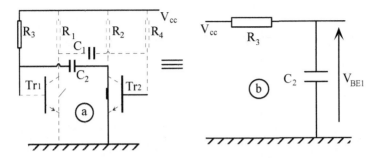

Figure 6.3. *Charge of capacitor C_2 when T_{r1} is blocked and T_{r2} is saturated*

Capacitor C_2 charges until voltage V_{BE1} exceeds the threshold voltage (about 0.6 V). At "$t = T_1$", transistor Tr1 becomes conductive and, by cumulative effect, it rapidly saturates. Therefore, $V_{CE1} \cong 0$.

Before this, we had $V_{CE1} \cong +V_{cc}$, which means a sudden variation of about $-V_{cc}$ (shift from $+V_{cc}$ to 0 V). This abrupt variation is instantaneously transmitted to the base of transistor Tr2 through capacitor C_1. It should be recalled that V_{BE2} was equal to 0.6 V. For the sake of simplification, let us consider $V_{BE2} \cong 0$ V. At $t = T_1$, voltage V_{BE2} has a variation of $-V_{cc}$. Then, $V_{BE2} \cong -V_{cc}$ (in fact, $V_{BE2} \cong -V_{cc} + 0.6$ V). Transistor Tr2 is instantaneously blocked. Starting from instant $t = T_1$, the astable circuit (Figure 6.4(a)) is equivalent to the diagram shown in Figure 6.4(b).

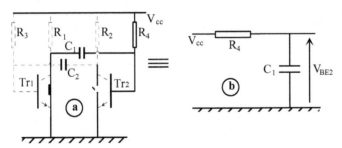

Figure 6.4. *a) State of the astable circuit and b) equivalent diagram of the astable circuit when transistor T_{r2} is blocked and transistor T_{r1} is saturated*

Under these conditions, capacitor C_1 charges through resistance R_4 starting from an initial voltage equal to approximately $-V_{cc}$. When the voltage across the capacitor reaches 0.6 V, transistor Tr2 conducts and rapidly reaches saturation. Its "collector–emitter" voltage suddenly shifts from $V_{CE2} = V_{cc}$ to $V_{CE2} = 0$.

This sudden variation ($-V_{cc}$) is instantaneously transmitted by capacitor C_2 to the base of transistor Tr1. Hence, $V_{BE1} = 0.6$ V ($\cong 0$ V) $\rightarrow V_{BE1} = -V_{cc} + 0.6$ V ($\cong -V_{cc}$).

Then, transistor Tr1 is blocked (Tr2 is saturated), and the circuit is in the state that is schematically shown in Figure 6.3. The cycle thus described repeats indefinitely, and a rectangular or square signal will be obtained at the output of the collector of Tr1 or Tr2. For a better understanding of the operation of astable device with transistors, the evolution of various signals present in bases and collectors of two transistors Tr1 and Tr2 is shown in Figure 6.5.

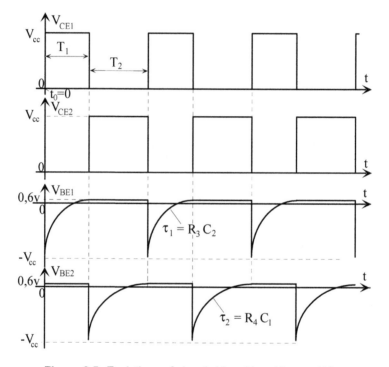

Figure 6.5. *Evolutions of signals V_{CE1}, V_{CE2}, V_{BE1} and V_{BE2}*

Choosing the output on the collector of transistor Tr1 yields:

$$v_s = V_{CE1} = \begin{cases} V_{cc} & t_0 < t < T_1 \\ 0 & T_1 < t < T_2 + T_1 \end{cases}$$

The initial state t_0 is chosen equal to zero ($t_0 = 0$).

6.2.5. *Period of the output signal*

In order to calculate the period of the output signal, it is important to establish the equations for the charge of capacitors C_1 and C_2. The duration of the high state of the output signal (e.g. on the collector of Tr1) is the time needed for the capacitor C_2 to charge from the initial value $(-V_{cc} + 0.6\ V)$ to the final value of $+ 0.6\ V$ (for the sake of simplicity, the initial voltage can be considered equal to $-V_{cc}$ and the final voltage is $0\ V$). The duration of the low state is the time needed for the charge of capacitor C_1 to pass from an initial voltage of $-V_{cc}$ to a final voltage of the order of $0\ V$.

6.2.5.1. *Calculation of the duration of the high state*

The following equation defines voltage V_{C2} of the charge of capacitor C_2:

$$V_{BE1} = V_{C2} = A.e^{\frac{-t}{\tau_1}} + B \quad \text{with} \quad \tau_1 = R_3 C_2$$

where "A" and "B" are two constants that are determined using boundary conditions and τ_1 is the charging time constant of capacitor C_2.

For $t = 0$, $V_{C2} = -V_{cc}$ and for $t \to \infty$, $V_{C2} \to V_{cc}$.

Furthermore:

$$A + B = -V_{cc} \text{ and } B = V_{cc}; \text{ hence, } A = -2V_{cc}$$

Under these conditions, the equation for the charge of C_2 can be written as follows:

$$V_{BE1} = V_{C2} = V_{cc}(1 - 2.e^{\frac{-t}{\tau_1}})$$

At $t = T_1$, the voltage across capacitor C_2 is $0\ V$:

$$V_{C2} = V_{cc}(1 - 2.e^{\frac{-T_1}{\tau_1}}) = 0$$

$$T_1 = R_3 C_2 . Ln(2) \cong 0.69.R_3 C_2$$

6.2.5.2. *Calculation of the duration of the low state*

The following is the equation of the charge of capacitor C_1:

$$V_{BE2} = V_{C1} = D.e^{\frac{-t}{\tau_2}} + E \quad \text{with} \quad \tau_2 = R_4 C_1$$

where constants (D and E) are determined using the boundary conditions: When t = 0, $V_{C1} = -V_{cc}$, and when t → ∞, $V_{C1} → V_{cc}$. This leads to D + E = $-V_{cc}$ and E = V_{cc}; hence, D = $-2V_{cc}$.

The equation of the charge of capacitor C_1 in its final form is:

$$V_{BE2} = V_{C1} = V_{cc}(1 - 2.e^{\frac{-t}{\tau_2}})$$

At t = T_2, voltage V_{C1} = 0:

$$V_{C1} = V_{cc}(1 - 2.e^{\frac{-T_2}{\tau_2}}) = 0$$

$$T_2 = R_4C_1.Ln(2) \cong 0.69.R_4C_1$$

The period of the output signal is: $T = T_1 + T_2 = 0.69(R_3C_2 + R_4C_1)$

NOTE.– In the general case, the signal is rectangular; it can be square when: $R_3C_2 = R_4C_1$. In order to obtain this result, it is sufficient to choose two variable resistances for R_3 and R_4, leading to the following equality:

$$\frac{R_4}{R_3} = \frac{C_2}{C_1}$$

6.3. Astable device with operational amplifier

An astable circuit can be readily realized using, for example, an operational amplifier, as shown in Figure 6.6.

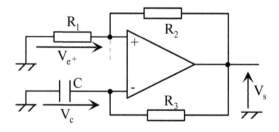

Figure 6.6. *Astable device with operational amplifier*

6.3.1. *Operating principle*

6.3.1.1. *First stage*

At the non-inverting input of the operational amplifier, there is a feedback voltage v_e+:

$$V_{e^+} = \frac{R_1}{R_1 + R_2} V_s$$

Due to the unstable state of the voltage across capacitor C (presence of a positive feedback due to the return chain ensured by R_1 and R_2), the output of the amplifier will be in a high or low state. Let us suppose that the output is in a high state: $v_s = V_{cc}$ where V_{cc} is the positive supply voltage of the operational amplifier. It should be recalled that the latter is symmetrically supplied ($\pm V_{cc}$) (in reality, the saturation voltage is slightly below the supply voltage due to the existence of a waste voltage related to operational amplifiers).

The voltage brought back by the operational amplifier at its positive input is:

$$V_1 = v_e^+ = \frac{R_1}{R_1 + R_2} V_{cc}$$

The capacitor C charges through resistance R_3 in order to try to reach the voltage V_{cc} applied to it. The law that governs the variations of voltage (V_c) across the capacitor can be defined by the following equation:

$$v_c = v_{e-} = A.e^{\frac{-t}{\tau}} + B \quad \text{with} \quad \tau = R_3 C$$

where "A" and "B" are constants that are determined using boundary conditions.

As long as the voltage v_e+ applied at the non-inverting input is above voltage v_e- applied at the inverting input, the output of the operational amplifier is in high state. On the contrary, if:

$$v_c = v_e- \geq V_1$$

then $v_s = -V_{cc}$

$$V_{e^+} = -\frac{R_1}{R_1 + R_2} V_{cc} = -V_1$$

6.3.1.2. *Second stage*

From the instant when the output voltage shifts from the high state to the low state, capacitor C starts to discharge across resistance R_3 in order to try to reach the voltage applied to it ($-V_{cc}$).

Nevertheless, as soon as the voltage across capacitor C reaches the value $-V_1$ and slightly exceeds it by a lower value, $v_{e^+} > v_{e^-}$ and $v_s = V_{cc}$

$$v_{e^+} = \frac{R_1}{R_1 + R_2} V_{cc} = V_1$$

The voltage now applied to capacitor C via resistance R_3 is equal to V_{cc}. Then, the capacitor starts to charge from the initial voltage ($-V_1$) in order to try to reach $+V_{cc}$.

This leads back to the cycle of the first stage, the second stage and so on.

The indefinite repetition of these cycles generates in the astable circuit the signals shown in Figure 6.7.

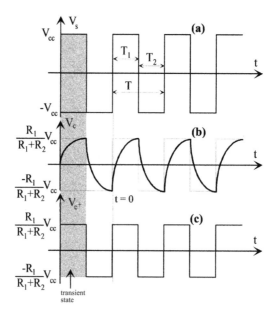

Figure 6.7. *Signals involved in the astable circuit with operational amplifier:*
a) output signal; b) signal across the capacitor; and c) signal at
the non-inverting input

6.3.2. *Period of the output signal*

The capacitor charges and discharges with the same time constant: $\tau = R_3.C$

As already noted, the capacitor charges according to the following law:

$$v_c = A.e^{\frac{-t}{\tau}} + B$$

where the constants "A" and "B" are determined by analyzing the evolution of voltage V_c schematically shown in Figure 6.7.

"A" and "B" are obtained for an established steady state. The initial instant $t = 0$ of this steady state is chosen as indicated in Figure 6.7.

At $t = 0$, $v_c = A + B$ and $V_c = \dfrac{-R_1}{R_1 + R_2} V_{cc}$;

when $t \rightarrow \infty$; $v_c = V_{cc}$

This yields:

$$B = V_{cc} \quad \text{and} \quad A = -\left[1 + \frac{R_1}{R_1 + R_2}\right] V_{cc}$$

$$v_c = V_{cc}\left[1 - \left[1 + \frac{R_1}{R_1 + R_2}\right] e^{\frac{-t}{R_3.C}}\right]$$

It should be noted that the period T is the sum of the durations of the high state and low state: $T = T_1 + T_2$.

Thus, given that the charge and discharge time constants of the capacitor are identical and that the shift voltages are symmetrical, it can be stated that the duration of the high state is equal to the duration of the low state, and the resulting signal is in this case square: $T_1 = T_2$ and $T = 2T_1$.

At $t = T_1$, the value of voltage V_c across the capacitor is equal to:

$$V_c = \frac{R_1}{R_1 + R_2} V_{cc} = V_{cc}\left[1 - \left[1 + \frac{R_1}{R_1 + R_2}\right] e^{\frac{-T_1}{R_3.C}}\right]$$

$$T_1 = R_3.C.Ln\left[1 + 2\frac{R_1}{R_2}\right]$$

$$T = 2R_3.C.Ln\left[1 + 2\frac{R_1}{R_2}\right]$$

6.4. Astable circuit with voltage-controlled frequency

6.4.1. Principle and operation

Let us go back to the astable multivibrator circuit with operational amplifier and slightly modify it in order to obtain the circuit shown in Figure 6.8. The frequency of the output signal depends on the application of an external voltage V_{ex}.

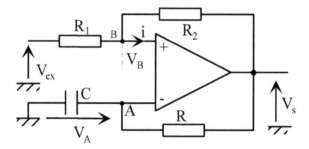

Figure 6.8. *Astable circuit with voltage-controlled frequency*

The output voltage of the circuit can only have two levels: a high-amplitude level ($+V_{cc}$) or a low-amplitude level ($-V_{cc}$). For a more simple explanation of how this device operates, let us suppose that the output is at high state: $v_s = +V_{cc}$.

The value of the differential input impedance of the operational amplifier is considered infinite. Consequently, the current "i" is zero. Applying, for example, Millman's theorem allows the deduction of the following relation:

$$V_B = V_1 = \frac{R_1 V_{cc} + R_2 V_{ex}}{R_1 + R_2}$$

The capacitor starts to charge across resistance R:

$$V_{Ach} = A_1 . e^{\frac{-t}{\tau}} + B_1 \quad \text{with} \quad \tau = RC$$

where A_1 and B_1 are constants to be determined using boundary conditions.

At the instant when voltage V_A across the capacitor slightly exceeds V_1, the output of the operational amplifier shifts from the high state to the low state: $v_s = -V_{cc}$:

$$V_B = V_2 = \frac{-R_1 V_{cc} + R_2 V_{ex}}{R_1 + R_2}$$

The capacitor then starts to discharge across resistance R according to the law:

$$V_{Adéc} = D . e^{\frac{-t}{\tau}} + E \quad \text{with} \quad \tau = RC$$

where "D" and "E" are constants that are determined using boundary conditions.

When the voltage (V_A) slightly exceeds the quantity V_2, the output of the astable circuit shifts once more, passing this time from the low state to the high state: $v_s = V_{cc}$

$$V_B = V_1 = \frac{R_1 V_{cc} + R_2 V_{ex}}{R_1 + R_2}$$

The capacitor restarts charging until the voltage across it reaches and slightly exceeds V_1.

The output will still shift and the cycles described will repeat infinitely in order to generate in the astable circuit the signals (v_S, V_A and V_B) schematically shown in Figure 6.9.

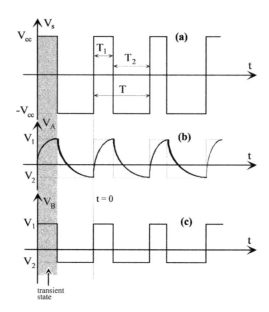

Figure 6.9. *Evolution of the signals involved in the astable circuit at variable frequency: a) output signal v_s; b) signal V_A across the capacitor and c) signal V_B at the non-inverting input*

6.4.2. *Period of the output signal*

The period of the output signal is defined as $T = T_1 + T_2$

where T_1 and T_2 are the durations of the high state and low state, respectively.

The calculation of T_1 involves finding the time required for the capacitor charge voltage to shift from an initial voltage equal to V_2 to a final voltage equal to V_1.

On the contrary, the duration T_2 relates to a discharge of this capacitor, whose voltage passes from an initial value equal to V_1 to a final value equal to V_2.

In order to find the expressions of the high-state and low-state durations, the relations that define the charge and discharge voltages of capacitor C should be found first.

6.4.2.1. Duration of the high state

It should be recalled that the capacitor charge voltage across resistance R is defined as:

$$V_{Ach} = A_1.e^{\frac{-t}{\tau}} + B_1$$

In order to calculate the duration T_1 of the high state of the output signal, the expressions that define constants A_1 and B_1 must be found.

At $t = 0$; $V_A = A_1 + B_1 = V_2$;

$$A_1 + B_1 = V_2 = \frac{-R_1 V_{cc} + R_2 V_{ex}}{R_1 + R_2}$$

When $t \to \infty$, $V_A = B_1 = V_{cc}$ and $A_1 = \frac{-2R_1 V_{cc} + R_2 (V_{ex} - V_{cc})}{R_1 + R_2}$;

Under these conditions, the equation of the charge voltage can be written as:

$$V_{Ach} = V_{cc} + \frac{-2R_1 V_{cc} + R_2 (V_{ex} - V_{cc})}{R_1 + R_2} e^{\frac{-t}{\tau}}$$

At $t = T_1$, the value of the charge voltage corresponds to V_1:

$$V_{Ach}(t = T_1) = V_{cc} + \frac{-2R_1 V_{cc} + R_2 (V_{ex} - V_{cc})}{R_1 + R_2} e^{\frac{-T_1}{\tau}} = V_1 = \frac{R_1 V_{cc} + R_2 V_{ex}}{R_1 + R_2}$$

$$T_1 = RC.Ln \left[\frac{V_{cc}(2R_1 + R_2) - R_2 V_{ex}}{R_2 (V_{cc} - V_{ex})} \right]$$

6.4.2.2. Duration of the low state

The expression of the capacitor discharge voltage is:

$$V_{Adec} = D.e^{\frac{-t}{\tau}} + E$$

For the calculation of the duration T_2 of the low state of the output signal, the expressions of constants D and E must be defined.

At t = 0, $V_{Adec} = D + E = V_1$; $D + E = V_1 = \dfrac{R_1 V_{cc} + R_2 V_{ex}}{R_1 + R_2}$,

When $t \to \infty$, $V_{Adec} = E = -V_{cc}$ and $D = \dfrac{2R_1 V_{cc} + R_2 (V_{ex} + V_{cc})}{R_1 + R_2}$

The expression of the discharge voltage in its final form is then defined as:

$$V_{Adec} = -V_{cc} + \frac{2R_1 V_{cc} + R_2 (V_{ex} + V_{cc})}{R_1 + R_2} e^{\frac{-t}{\tau}}$$

At t = T_2, the value of the charge voltage corresponds to V_2.

$$V_{Adec}(t = T_2) = -V_{cc} + \frac{2R_1 V_{cc} + R_2 (V_{ex} + V_{cc})}{R_1 + R_2} e^{\frac{-T_2}{\tau}}$$

$$V_{Adec}(t = T_2) = V_2 = \frac{-R_1 V_{cc} + R_2 V_{ex}}{R_1 + R_2}$$

$$T_2 = RCLn \left[\frac{V_{cc}(2R_1 + R_2) + R_2 V_{ex}}{R_2 (V_{cc} + V_{ex})} \right]$$

6.4.2.3. *Period of the output signal*

As previously written, the period T is defined as:

T = T_1 + T_2

$$T = RCLn \left[\frac{\left[\left[1 + \dfrac{2R_1}{R_2} \right] \right]^2 - \left[\dfrac{V_{ex}}{V_{cc}} \right]^2}{1 - \left[\dfrac{V_{ex}}{V_{cc}} \right]^2} \right]$$

The variations of period T as a function of V_{ex} are schematically shown in Figure 6.10.

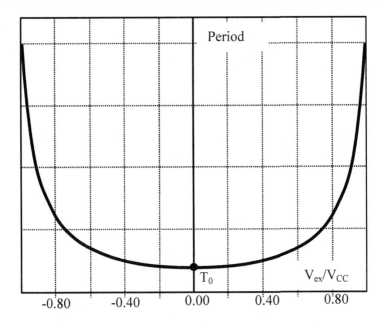

Figure 6.10. *Evolution of the period of the output signal with external voltage* V_{ex}

For $V_{ex} = 0$, a similar result to the one already established in Figure 6.3 for the astable circuit with operational amplifier in the absence of external voltage is obtained:

$$T_0 = 2RCLn\left[1+(\frac{2R_1}{R_2})\right]$$

6.5. Timer-based astable circuit (555 integrated circuit)

6.5.1. Presentation of 555 timer

The simplified internal diagram of a 555 circuit is shown in Figure 6.11.

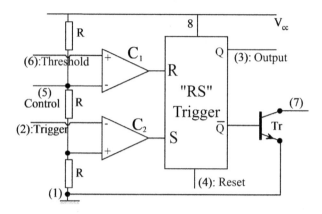

Figure 6.11. *Simplified internal diagram of a 555 integrated circuit*

The "DIL" (dual in line) configuration of the 555 package and its pinning are shown in Figure 6.12 and in Table 6.1.

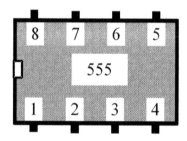

Figure 6.12. *Package of the 555 integrated circuit*

Pin No.	Function
1	Ground
2	Trigger
3	Output
4	Reset
5	Control
6	Threshold
7	Discharge
8	Supply

Table 6.1. *Package of the 555 integrated circuit*

6.5.2. *Operating principle of the 555-based astable circuit*

This circuit is essentially composed of a resistive network of three equal resistances, two comparators, a set-reset (SR) trigger and a bipolar transistor. The resistive network allows the division of the supply voltage into two voltages $(1/3)V_{cc}$ and $(2/3) V_{cc}$ that will serve as reference voltages to the two comparators C_1 and C_2. The outputs of these two circuits control the inputs R (reset) and S (set) of the SR trigger.

The output of the SR trigger is the output of the 555 timer. The complementary output of the SR trigger serves for the operation in switching the mode of the transistor Tr (called the discharge transistor).

In order to realize a circuit that generates rectangular waves with a 555 timer, it is sufficient to have few external components, as shown in Figure 6.13.

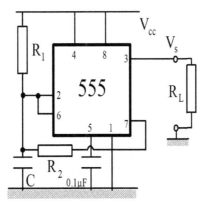

Figure 6.13. *Astable device realized with a 555 integrated circuit*

Initially, the potential of the positive input of the comparator C_2 is equal to $(1/3)V_{cc}$, whereas the potential of its negative input is zero (the capacitor is obviously supposed to be discharged when operation starts). The potential $(2/3)V_{cc}$ at the negative input of comparator C_1 is higher than the potential applied at the positive input. Consequently, the inputs and the output of SR triggers are in the following states:

$$\left.\begin{array}{l} R = 0 \\ S = 1 \end{array}\right\} \Rightarrow Q = 1 \quad \text{and} \quad \overline{Q} = 0 ; \qquad\qquad v_s \cong V_{cc}$$

The capacitor C starts to charge, seeking to reach the voltage $(+V_{cc})$ applied to it. When the voltage across it reaches and slightly exceeds $(1/3)V_{cc}$, the following states are reached for the inputs and output of the SR trigger:

$$\left.\begin{array}{l} R = 0 \\ S = 0 \end{array}\right\} \Rightarrow Q = 1 \quad \text{and} \quad \overline{Q} = 0$$

The previous state for the output of SR trigger has been maintained.

The capacitor charge voltage continues to evolve, and when it reaches and slightly exceeds $(2/3)V_{cc}$, the following states are reached for the two inputs and the output of the SR trigger:

$$\left.\begin{array}{l} R = 1 \\ S = 0 \end{array}\right\} \Rightarrow Q = 0 \quad \text{and} \quad \overline{Q} = 1 \, ; \qquad\qquad v_s = 0$$

The discharge transistor is saturated under the influence of the voltage applied to its base: $V_B \cong V_{cc}$ ($\overline{Q} = 1$). Consequently, the potential of pin (7) of the 555 circuit is zero.

Therefore, capacitor C starts to discharge through resistance R_2 until the voltage across it reaches and exceeds $(1/3)V_{cc}$ by lower values. At this instant, the states of the inputs and output of the SR trigger become:

$$\left.\begin{array}{l} R = 0 \\ S = 1 \end{array}\right\} \Rightarrow Q = 1 \quad \text{and} \quad \overline{Q} = 0 \, ; \qquad\qquad v_s \cong V_{cc}$$

When the voltage applied to its base is zero ($\overline{Q} = 0$), the discharge transistor is blocked and pin (7) becomes an open connection.

The capacitor has no possibility to discharge. It starts its charge through resistance R_1 until the voltage across it reaches and slightly exceeds by a $(2/3)V_{cc}$ higher value. At this moment, the inputs and the output have the following states:

$$\left.\begin{array}{l} R = 1 \\ S = 0 \end{array}\right\} \Rightarrow Q = 0 \quad \text{and} \quad \overline{Q} = 1$$

The described cycles are indefinitely repeated. The shapes of the signals obtained across the capacitor and at the output are schematically shown in Figure 6.14.

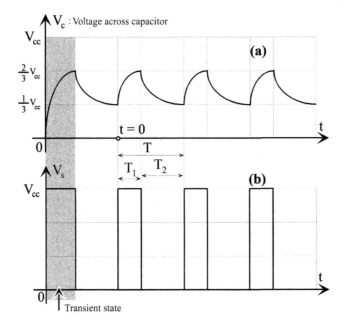

Figure 6.14. *Evolution of a) the signal across the capacitor and b) the output signal of the 555 astable circuit*

NOTE.– To ensure proper operation of the astable circuit, it is imperative that:

$R_1 > 2R_2$.

This condition allows the capacitor discharge voltage to reach and even go below the threshold voltage value of $(1/3)V_{cc}$. Otherwise, the circuit cannot operate, because the low threshold voltage cannot be reached. If $R_1 \leq 2R_2$, then there is no instability phenomenon.

6.5.3. *Period of the output signal*

In order to calculate the period T of the output signal, it is important to calculate the duration T_1 of the high state and the duration T_2 of the low state of this signal:

$T = T_1 + T_2$

6.5.3.1. *Duration of the high state*

Let us assume that the astable circuit has reached its steady state. The initial instant is chosen at t = 0 (established steady state) for the calculation of T_1 (duration of the high state). The analysis of the shape of the signals involved (see the diagram in Figure 6.14) evidences that the duration of the high state is the time needed by the capacitor C to charge from an initial voltage of $(1/3)V_{cc}$ to a final voltage of $(2/3)V_{cc}$. A capacitor charge voltage is defined by the following general relation:

$$V_c = V_{ch\,arg\,e} = A.e^{\frac{-t}{R_1C}} + B$$

where constants "A" and "B" are determined using boundary conditions.

At t = 0; $V_c = (1/3)V_{cc}$. When t → ∞, $V_c = V_{cc}$.

This yields: $A + B = \frac{1}{3}V_{cc}$; $B = V_{cc}$ and $A = \frac{-2}{3}V_{cc}$

$$V_c = V_{ch\,arg\,e} = V_{cc}(1 - \frac{2}{3}e^{\frac{-t}{R_1C}})$$

At t = T_1, the charge voltage across the capacitor is equal to $(2/3)V_{cc}$.

$$V_c(t = T_1) = V_{cc}(1 - \frac{2}{3}e^{\frac{-T_1}{R_1C}}) = \frac{2}{3}V_{cc}$$

$$T_1 = R_1.C.Ln(2) \cong 0.69.R_1.C$$

6.5.3.2. *Duration of the low state*

The duration of the low state is the time needed by the capacitor to discharge from an initial voltage of $(2/3)V_{cc}$ to a final voltage of $(1/3)V_{cc}$. In order to find the expression that defines the duration of the low state, it is important to first find the law of the variation of discharge voltage across capacitor C:

$$V_c = V_{disch} = D.e^{\frac{-t}{R_2C}} + E$$

where constants "D" and "E" are determined as follows: at t = 0 (for the discharge), $V_c = D + E = (2/3)V_{cc}$; when t → ∞, $V_c = E = 0$. Then:

$$D = (2/3)V_{cc}.$$

Finally, the discharge voltage is expressed as:

$$V_c = V_{disch} = \frac{2}{3}V_{cc}.e^{\frac{-t}{R_2C}}$$

At t = T_2, the value of the discharge voltage is equal to $(1/3)V_{cc}$:

$$V_c(t=T_2)=V_{disch}(t=T_2)=\frac{2}{3}V_{cc}.e^{\frac{-T_2}{R_2C}}=\frac{1}{3}V_{cc}$$

Hence, the expression of T_2 (duration of the low state) is:

$$T_2 = R_2.C.Ln(2) \cong 0.69.R_2.C$$

Finally, the period of the output signal is defined by:

$$T = (R_2.C + R_1.C)Ln(2) \cong 0.69(R_1.C + R_2.C)$$

6.5.4. *Another possible implementation*

Astable circuits can be realized by changing the arrangement of the external elements with respect to the circuit indicated in section 6.5.2. Figure 6.15 provides an example. Further implementations are possible.

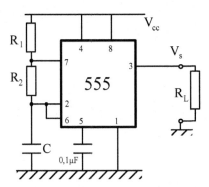

Figure 6.15. *Another implementation of the design of a (555) astable timer*

The capacitor charges through resistances R_1 and R_2 and discharges only through resistance R_2. Consequently, the durations of the high state (T_1) and low state (T_2) are expressed as:

$$T_1 = (R_1 + R_2)C.\text{Ln}(2)$$

$$T_2 = R_2.C.\text{Ln}(2)$$

The period T is:

$$T = (R_1 + 2R_2)C.\text{Ln}(2)$$

NOTE.– It is possible to obtain an output signal of variable frequency. This can be achieved by replacing one of the two resistances R_1 or R_2 by a potentiometer. It is also possible to vary this frequency automatically by injecting a variable signal in the control command pin (5) of the 555 circuit.

6.6. Astable multivibrators with logic gates

6.6.1. *Principle and operation*

Logic gates offer a very simple way to realize astable circuits. An example of an astable multivibrator designed with "NOT AND" or NAND logic gates is shown in Figure 6.16.

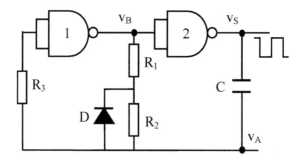

Figure 6.16. *Astable multivibrator with "NOT AND" logic gate*

The output voltage is initially zero: $v_s = 0$. Under these conditions, $V_B = V_{cc}$ (V_{cc}: supply voltage) and consequently $V_A = 0$. Capacitor C charges through resistances R_1 and R_2 according to the diagram shown in Figure 6.17. Diode D is blocked.

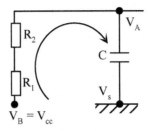

Figure 6.17. *Charge of capacitor C through (R₁ + R₂). The output potential is 0 V*

For a more simple description of the operation, let us suppose that the switching threshold of the gates is ($V_{cc}/2$). When the capacitor starts to charge, the potential of point A tends to increase. When the potential at this point reaches and slightly exceeds the switching threshold ($V_{cc}/2$), voltages at point B and at the output become $V_B = 0$ and $v_s = V_{cc}$, respectively.

The sudden variation (from 0 to $+V_{cc}$) of voltage v_s is instantaneously transmitted from the output to point A through capacitor C.

Consequently, $V_A = \dfrac{3}{2}V_{cc}$, $V_B = 0$ and $v_s = V_{cc}$

Voltage V_A is directly applied at the input of the first "NOT AND" gate and on the anode of diode D. The diode is then conducting. It short-circuits resistance R_2. At this precise instant, the astable multivibrator with "NOT AND" gates can be represented by the equivalent diagram shown in Figure 6.18.

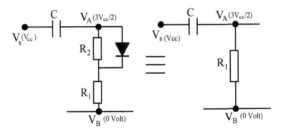

Figure 6.18. *State of the astable circuit with logic gates when $v_A = \dfrac{3}{2}V_{cc}$; $V_B = 0$ and $v_s = V_{cc}$*

The capacitor charges, and the voltage across it is maximal. The voltage (V_A) across resistance R_1 tends to decrease in the same rate as that of the capacitor charge. The evolution of voltage V_A obeys the following law:

$$V_{R_1} = V_A = De^{\frac{-t}{R_1C}} + E$$

where constants "D" and "E" are determined as follows:

At t = 0, V_A = (3V_{cc}/2), and when t → ∞, V_A = 0 (there is no current flow through resistance R_1). E = 0;   and $D = \dfrac{3}{2}V_{cc}$

$$V_{R_1} = V_A = \frac{3}{2}V_{cc}.e^{\frac{-t}{R_1C}}$$

The variation of the voltage across the capacitor is defined as:

$$V_c = V_{cc} - V_{R_1}$$

$$V_c = V_{cc}(1 - \frac{3}{2}e^{\frac{-t}{R_1C}})$$

The voltage at point "A" decreases with time until it reaches $V_{cc}/2$ and slightly exceeds it. At this instant, $V_B = V_{cc}$ and $v_s = 0$

The output voltage suddenly decreases from V_{cc} to 0 V. This sudden diminution is instantaneously transmitted from the output to point "A" through capacitor C:

$$V_B = V_{cc} \text{ and } V_s = 0$$

$$V_A = \frac{V_{cc}}{2} - V_{cc} = -\frac{V_{cc}}{2}$$

Under these conditions, the operating state of the astable circuit with logic gates can be summarized by the diagram shown in Figure 6.19. It can be noted that this diagram is practically identical to the one shown in Figure 6.17.

The only difference is the initial voltage at point "A". In this case, it is $(-V_{cc}/2)$, while initially it was zero (transient state).

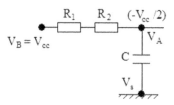

Figure 6.19. *Astable circuit with logic gates in the case where $V_B = V_{cc}$, $v_s = 0$ and $V_A = \dfrac{-V_{cc}}{2}$*

From this instant, the steady state is established and the previously described cycle repeats indefinitely as long as the circuit is powered. The various signals existing in the astable circuit with logic gates are summarized in Figure 6.20.

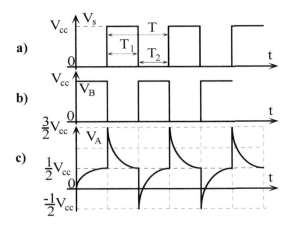

Figure 6.20. *Signals involved in the astable circuit with logic gates:*
a) output signal; b) signal at point B; and c) signal at point A

6.6.2. *Period of the output signal*

The period of the output signal is the sum of the duration T_1 of the high state and the duration T_2 of the low state:

$$T = T_1 + T_2$$

Therefore, in order to find the period, it is sufficient to express the two quantities T_1 and T_2.

6.6.2.1. *Duration of the high state*

The calculation of the duration of the high state T_1 involves finding the time needed for voltage at point A to pass from an initial value of $3V_{cc}/2$ to a final value of $V_{cc}/2$. It should be recalled that the law of variation of V_A (when $v_s = V_{cc}$) is defined as:

$$V_{R_1} = V_A = \frac{3}{2} V_{cc} \cdot e^{\frac{-t}{R_1 C}}$$

At t = 0, $V_A = (3V_{cc}/2)$. At t = T_1, $V_A = (V_{cc}/2)$.

$$V_A(t = T_1) = \frac{3}{2}V_{cc}.e^{\frac{-T_1}{R_1 C}} = \frac{V_{cc}}{2}$$

$T_1 = R_1.C.Ln(3)$

6.6.2.2. Duration of the low state

The duration T_2 of the low state is the time needed by voltage at point "A" to pass from a value $(-V_{cc}/2)$ to a value $(V_{cc}/2)$. The output voltage is zero throughout the duration T_2. The law that governs the variation of voltage at point "A" is defined by the following expression:

$$V_A = V_{cc}(1 - \frac{3}{2}e^{\frac{-t}{(R_1+R_2)C}})$$

At t = 0, $V_A = (-V_{cc}/2)$. At t = T_2, the value of voltage V_A is equal to $V_{cc}/2$. In this case, the capacitor charging constant is equal to $(R_1 + R_2)C$:

$$V_A = V_{cc}(1 - \frac{3}{2}e^{\frac{-T_2}{(R_1+R_2)C}}) = \frac{V_{cc}}{2}$$

$T_2 = (R_1 + R_2).C.Ln(3)$

The period is then defined by the following expression:

$T = (2R_1 + R_2).C.Ln(3)$

NOTE.– The technology used for the logic gates relies essentially on the expected outputs and on the use frequency. Thus, when the frequency parameter is relatively low and quite significant logic levels are expected in terms of voltage, it is preferable to use "MOS" logic gates. When a compromise must be reached between the output level and the use frequency, it is advisable to use "TTL" gates. When the output frequency parameter is expected to be high, "ECL" technology should be the choice.

M.O.S: Metal-oxide semiconductor

T.T.L: Transistor–transistor logic (use of bipolar transistors in switching mode)

E.C.L: Emitter-coupled logic (this technology employs bipolar transistors with emitter coupling. The implementation allows a very rapid switch due to the fact that transistors never reach saturation).

6.7. Astable multivibrators with specialized integrated circuits

6.7.1. *Introduction*

As noted above, a rectangular or square wave can be generated in various ways. In electronics, integration has further opened the door to specialized circuits, in order to fulfill a quite precise function. Some of these circuits are described in the following.

6.7.2. *Specification of the 74123 integrated circuit*

The 74123 integrated circuit (Figure 6.21) uses T.T.L. technology. It is marketed as a "DIL" (dual in line) package with 16 pins. Its internal configuration is essentially composed of two monostables.

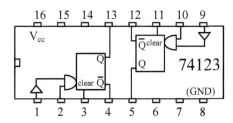

Figure 6.21. *Diagram of the 74123 integrated circuit*

The pinning of this circuit is indicated in Table 6.2.

Pin	Function
1; 9	Inputs 1_A and 2_A
2; 10	Inputs 1_B and 2_B
3; 11	Inputs "clear1" and "clear2"
4; 12	Complementary 1 and 2
5; 13	Outputs Q_2 and Q_1
6; 7 with 14 and 15	Connection of external components
8	Ground
16	Supply voltage V_{cc}

Table 6.2. *Pinning of 74123*

The simultaneous use of these two integrated monostables allows realization of an astable circuit. First, an overview of these two monostables will be given. To be implemented, the monostable needs a few external components, such as a capacitor and a resistance (Figure 6.22).

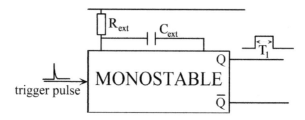

Figure 6.22. *Operating principle of a monostable circuit*

For the case that is of interest here (shown in Figure 6.22), it should be noted that when a pulse arrives at the input of the monostable circuit, the circuit output undergoes a transition from a stable state (low state) to a quasi-stable state (high state).

The duration T_1 of this quasi-stable high state is set by the external components R_{ext} and C_{ext}. This duration is generally defined using the following relation:

$$T_1 = R_{ext}.C_{ext}.Ln(2)$$

As soon as the duration T_1 expires, the monostable circuit returns to its stable state. The signals implemented in the monostable circuit are schematically shown in Figure 6.23.

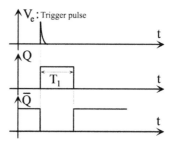

Figure 6.23. *Signals implemented in the monostable circuit*

6.7.3. *Astable circuit based on a 74123 integrated circuit*

The schematic diagram of an astable circuit, the design of which relies on a 74123 "TTL" integrated circuit, is shown in Figure 6.24.

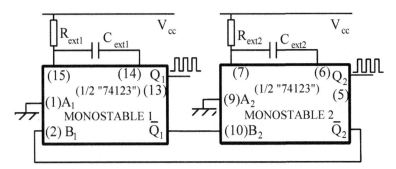

Figure 6.24. *Schematic diagram of the astable circuit with a 74123 integrated circuit*

The two "clear" inputs of the two monostables that constitute the 74123 integrated circuit must be brought to a high logic state: **"clear = 1"**.

The following is the truth table (Table 6.3) that governs the operation of the two monostables that constitute 74123.

Inputs of one of the monostables		Corresponding output	
A	B	Q	$\overline{Q}$
1	0	0	1
1	1	0	1
1	0	0	1
0	1	0	1
0	↑	⎍	⎍
↓	1	⎍	⎍

Table 6.3. *Truth table for 74123 integrated circuit*

As this truth table indicates, it can be noted that in order to obtain a high level at output, it is sufficient to maintain the input "A" at zero and to send a rising edge on the input "B" (positive pulse). The other solution is to maintain the input B at "1" and send a falling edge (or a negative pulse) on the input A. The first configuration (see the schematic diagram of the astable in Figure 6.24) is chosen in this case. In order to explain the operation of the astable, let us *a priori* suppose that the output of the first configuration (1/2 "74123") is in high state. This output can maintain this state (quasi-stable state of the monostable) only for a well-determined period set by the components C_{ext1} and R_{ext1}.

At the moment, when this output undergoes a transition from the high state to the low state, the complementary output $\overline{Q}_1$ shifts from the low state to the high state. Therefore, there is a rising edge at the input B_2 of the second configuration (1/2 "74123"). Consequently, the output $\mathbf{Q_2}$ is brought from the low state to the high state. The duration T_2 of this high state is set by the external components C_{ext2} and R_{ext2}. After this duration, the output $\mathbf{Q_2}$ shifts to the low state and the complementary output $\overline{Q}_2$ shifts from the low state to the high state. Therefore, a rising edge is applied at the input B_1 of the first configuration (1/2 "74123"). Then, the output $\mathbf{Q_1}$ shifts from the low state to the high state and the cycle thus described repeats infinitely. A rectangular signal is thus obtained on one of the outputs $\mathbf{Q_1}$ or $\mathbf{Q_2}$ (Figure 6.25). The period of this signal is set by the external components C_{ext1}, R_{ext1}, C_{ext2} and R_{ext2}.

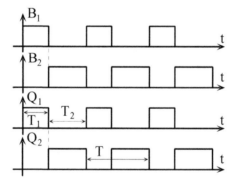

Figure 6.25. *Various signals implemented in the astable based on the 74123 integrated circuit*

The period of the output signal can be approximated by the following relation:

$$T = T_1 + T_2 = (C_{ext1}.R_{ext1} + C_{ext2}.R_{ext2}).Ln(2)$$

6.7.4. *Other specialized circuits*

6.7.4.1. *4047 integrated circuit*

This circuit is obtained by the "CMOS" manufacturing process; it can operate as a monostable and astable circuit. Its simplified internal block diagram is shown in Figure 6.26.

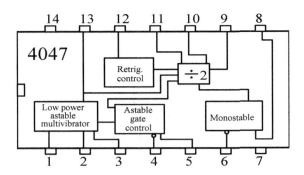

Figure 6.26. *Internal block diagram of 4047*
astable/monostable multivibrator integrated circuit

The pinning of 4047 is shown in Table 6.4.

Pinning	Function
1–3	C_x: external capacitor
2–3	R_x: external resistance
4–5	Astable operating command
6–8	Monostable operating command
7	Ground (GND) or supply voltage V_{SS}
9	Ext. reset
10	Output of monostable
11	Complementary output of monostable
12	Retrigger
13	Output of astable
14	Supply V_{CC}

Table 6.4. *Pinning of 4047*

6.7.4.2. *4528 integrated circuit*

This circuit is obtained by the "CMOS" manufacturing process. This integrated circuit comprises two monostables.

The realization of an astable based on this type of circuit has already been studied (see section 6.7.3). The internal block diagram of this circuit is shown in Figure 6.27.

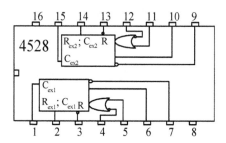

Figure 6.27. *Internal block diagram of the 4528 double monostable integrated circuit*

The functionalities of various pins of the 4528 CMOS integrated circuit are summarized in Table 6.5.

Pin	Function
1–2	C_{ext1}: external capacitor for monostable 1
2	R_{ext1}: external resistance for monostable 1
3	Reset 1 (monostable 1)
4–5	Control inputs of monostable 1 ↑ or ↓
6	Output of monostable 1
7	Complementary output of monostable 1
8	Ground (GND) or V_{SS}
9	Complementary output of monostable 2
10	Output of monostable 2
11–12	Control inputs of monostable 2 ↑ or ↓
13	Reset 2 (monostable 2)
14–15	C_{ext2}: external capacitor for monostable 2
14	R_{ext2}: external resistance for monostable 2
16	Supply V_{cc}

Table 6.5. *4528 pinning*

6.8. Exercises

EXERCISE 1

Let us consider the circuit shown in Figure E1.1. The capacitors are in each case assumed discharged at the start.

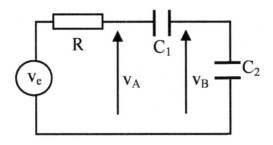

Figure E1.1.

1) Given are $v_e = E = 5$ V, $R = 1$ kΩ. Find the expressions of voltages v_A and v_B as a function of time (indicating each time the time constant) in the following cases:

1.1) $C_1 = 1$ μF and $C_2 = 1$ nF; 1.2) $C_1 = 1$ nF and $C_2 = 1$ μF;

1.3) $C_1 = C_2 = C = 1$ μF.

2) Draw the evolutions of v_A and v_B as a function of time; for the three cases in 1).

3) Let us now suppose that v_e is a unidirectional square signal (Figure E1.2) of amplitude 5 V and frequency 1,000 Hz.

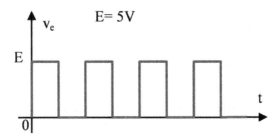

Figure E1.2.

Draw the representation in steady state of v_A and v_B as a function of time in the following cases (indicating for each the time constant):

3.1) $C_1 = 1$ µF and $C_2 = 1$ nF; 3.2) $C_1 = 1$ nF and $C_2 = 1$ µF;

3.3) $C_1 = C_2 = C = 1$ µF.

EXERCISE 2

Let us consider the circuit shown in Figure E2.1. As a first stage, the continuous voltage $v_e = E$ is applied (Figure E2.2) at the input. The capacitor is assumed discharged at the initial instant.

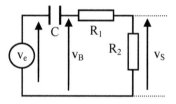

Figure E2.1.

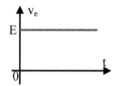

Figure E2.2.

1) Find the expressions and representations of v_s and v_B under the following conditions:

1.1) $R_1 = 10$ Ω; $R_2 = 100$ kΩ and $C = 1$ µF,

1.2) $R_1 = 100$ kΩ; $R_2 = 10$ Ω and $C = 1$ µF and

1.3) $R_1 = R_2 = R = 100$ kΩ and $C = 1$ µF.

2) This time, the signal v_e is a unidirectional square signal (Figure E2.3) of frequency $f = 10$ kHz and amplitude E. Write the expressions of voltages v_s and v_B for an established steady state under the following conditions:

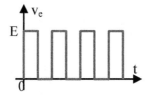

Figure E2.3.

2.1) $R_1 = 10\ \Omega$; $R_2 = 100\ k\Omega$ and $C = 1\ \mu F$,

2.2) $R_1 = 100\ k\Omega$; $R_2 = 10\ \Omega$ and $C = 1\ \mu F$ and

2.3) $R_1 = R_2 = R = 100\ \Omega$ and $C = 0.1\ \mu F$.

3) Draw the representation of signals v_e, v_s, v_B as a function of time for the three cases mentioned in question 2.

EXERCISE 3

The task is to study the behavior of the circuit shown in Figure E3.1 by applying at its input the square signal schematically shown in Figure E3.2.

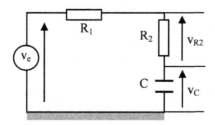

Figure E3.1.

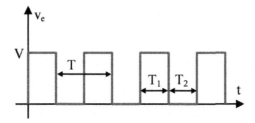

Figure E3.2.

1) Knowing that the time constant $\tau_1 = R_2C$ is very large compared to the period T of the signal applied at input, draw the representation of the evolution in time of the signals across resistances R_1 and R_2 and across capacitor C. For the sake of simplicity, a choice is made for $R_1 = R_2$.

2) Knowing that the time constant $\tau_2 = (R_1 + R_2)C$ is this time very small compared to the period T of the input signal, draw the evolution of signals across resistances R_1 and R_2 and across capacitor C. Describe the function of the circuit when the output is taken across resistance R_2.

EXERCISE 4

Let us consider the circuit shown in Figure E4.1. The voltage v_e is a square signal of frequency F = 1 kHz and amplitude E (Figure E4.2).

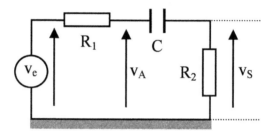

Figure E4.1.

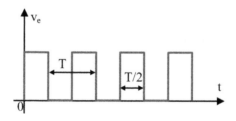

Figure E4.2.

NOTE.– Only the steady state should be considered throughout the exercise.

1) Express the charge and discharge constants of capacitor C and compare them to the half-period of the input signal in the following two cases: (a) $R_1 = R_2 = 1$ kΩ, C = 25 nF and (b) $R_1 = R_2 = 10$ kΩ, C = 20 μF. Draw the conclusions in each case.

2) $R_1 = R_2 = 1$ kΩ and $C = 25$ nF. Find the expressions of voltages v_{R1}, v_A and v_S and the voltage across capacitor C. Draw the graphic representation of these voltages one under the other.

3) $R_1 = R_2 = 10$ kΩ and $C = 20$ µF. Find the expressions of voltages v_{R1}, v_A and v_S and the voltage across capacitor C. Draw the graphic representation of these voltages one under the other.

EXERCISE 5

Let us consider the circuit shown in Figure E5.1, where $C_1 = C_2 = C = 1$ µF and $R = 1$ kΩ. The capacitors are discharged at $t = 0$.

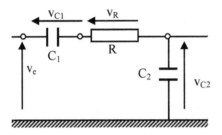

Figure E5.1.

1) Voltage v_e is a continuous voltage of amplitude 5 V. Find the expressions of v_{C1}, v_{C2} and v_R.

2) Draw these voltages as a function of time and find their final value.

3) This time v_e is a unidirectional square voltage of amplitude $E = 5$ V and frequency 100 Hz. Find the expressions of v_{C1}, v_R and v_{C2}. Draw the evolution of these voltages as a function of time.

EXERCISE 6

The circuit under study is schematically shown in Figure E6.1.

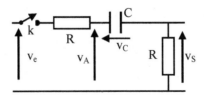

Figure E6.1.

1) $v_e = E = 5$ V, the switch k is closed at $t = 0$, with $v_C(0) = 0$. Find the expressions of v_S and v_A at $t = 0$ and throughout time.

2) Draw the evolution of v_S and v_A as a function of time.

3) This time, the signal applied at the input is a unidirectional square signal (varying from 0 V to E = 5 V) of frequency $f = 100$ Hz.

The switch is closed at $t = 0$ with $v_C(0) = 0$. Find the expressions of the charge and discharge time constants of the output voltage v_S and of the voltage at point A (v_A) for R = 1 kΩ and C = 0.1 μF.

4) Draw the evolution of v_S and v_A in time.

5) This time, R = 100 kΩ and C = 0.5 μF. Answer the same questions formulated at 3) and 4). Then, deduce the function achieved by the circuit studied at 5). Only the steady state case is taken into consideration.

EXERCISE 7

Let us consider the signal (Figure E7.1) applied at the input of a circuit (Figure E7.2).

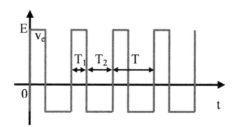

Figure E7.1.

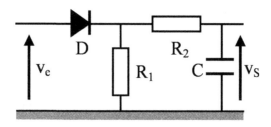

Figure E7.2.

$T_1 = 1$ ms; $T_2 = 2$ ms; $R_2 = 5$ $R_1 = 50$ kΩ; $C = 50$ nF and $E = 10$ V

1) The diode being assumed ideal, find the expressions and the graphic representation as a function of time of signals v_{R1} and v_s.

2) Draw the evolution in time of v_{R2}.

EXERCISE 8

Let us consider the circuit shown in Figure E8.1. Initially, the output is assumed to be in a high saturation state. The diode is considered ideal.

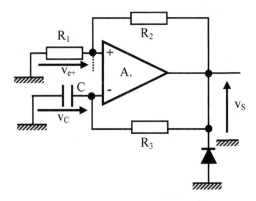

Figure E8.1.

1) Write the expressions of voltages v_{S-}, v_{e+} and v_C at this moment.

2) Since voltage v_{e-} evolves, find its value starting from which the output voltage switches. Under these conditions, find the expression and value of v_{e+}, v_C and v_S right after the switching. Draw the evolutions of these voltages. It is assumed that $R_1 = R_2 = R_3 = R$ for questions 2, 3 and 4.

3) Let us consider this time that the threshold of the diode is $V_0 = 0.6$ V. Rewrite the expression of various voltages (v_{e+}, v_C and v_S) and draw their evolution in time.

4) What type of signal is obtained at the output and what is its frequency?

EXERCISE 9

Let us consider the circuit shown in Figure E9.1, based on the 555 integrated circuit, the internal diagram of which is shown in Figure E9.2.

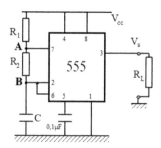

Figure E9.1.

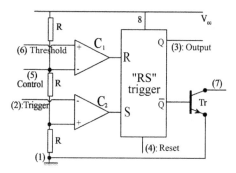

Figure E9.2.

1) Resume the diagram shown in Figure E9.1 and indicate the charge and discharge paths of capacitor C.

2) Briefly explain the operation of the circuit shown in Figure E9.1, relying on the internal structure of the integrated circuit and indicate the exact function of this circuit.

3) Draw the evolutions, one under the other, of various signals at points A and B and at the output.

4) Find the expressions and values of the output frequency "f" and of the cyclic ratio. Given are $R_1 = 10$ kΩ, $R_2 = 2$ kΩ and $C = 4.7$ nF.

EXERCISE 10

The circuit to be studied is represented by the diagram shown in Figure E10.1. At $t = 0$, the capacitor C is assumed discharged, the diode has zero threshold and the output voltage is $v_S = +V_{CC}$. Given: $V_{CC} = 14$ V.

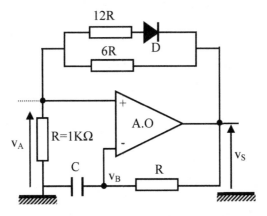

Figure E10.1.

1) Determine the state of diode D and the expressions and the values of v_A and of v_B at "$t = 0$".

2) Find the expressions of the charge and discharge time constants of capacitor C.

3) Find the expressions and the values of the switching thresholds.

4) Draw the evolutions of signals v_A, v_B and v_S and find the function of this circuit.

EXERCISE 11

Let us study the circuit shown in Figure E11.1. The NAND gates employ TTL technology ($V_{cc} = 5$ V).

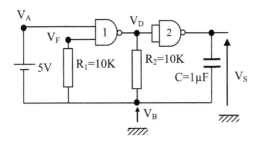

Figure E11.1.

1) Write the truth table of the NAND gate.

2) Initially, the output voltage v_S is assumed equal to $+V_{CC}$ and capacitor C is considered fully discharged. The shift threshold of the gates is $(V_{CC}/3)$. Study the operation of this circuit by drawing the evolutions of voltages v_B, v_D, v_F and v_S and of the voltage across capacitor C (v_C).

3) Assuming that steady state has been reached, find the expressions of v_A, v_B and v_C in the following cases:

3.1) v_S is at high state $(v_S = V_{CC})$;

3.2) v_S is at low state $(v_S = 0)$.

4) What is the function of this circuit? What happens if $R_1 = 0$?

5) What is the period T of the output signal (T_1 is the duration of the high state and T_2 is the duration of the low state)? Find the expression and the value of the cyclic ratio. What is the nature of the output signal?

EXERCISE 12

Let us consider the circuit shown in Figure E12.1. It features "NOR" gates, employing the TTL technology ($V_{CC} = 5$ V).

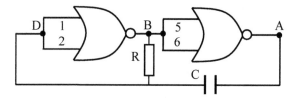

Figure E12.1.

1) Establish the truth table of a NOR gate with short-circuited inputs.

2) Provide a step-by-step explanation of the operation of this circuit, considering that initially $v_A = 0$ V. Then, deduce the evolution of voltages at points A, B and D.

3) The output is taken at point A. What is the function realized by this circuit? Calculate the duration of the high state and of the low state. Then, deduce the oscillation frequency. Numerical application: $R = 10$ kΩ and $C = 1$ μF

NOTE.– The switching threshold of the gates is assumed to be equal to $(V_{CC}/3)$.

EXERCISE 13[1]

The task is to study the circuit schematically shown in Figure E13.1, which employs a 555 timer.

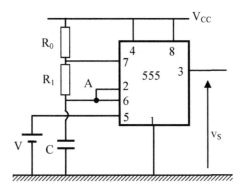

Figure E13.1.

1) What function does this circuit perform? Explain.

2) Represent one under the other the evolution in time of the signals at point A and at the output. Find their expression as a function of V and of parameters external to the "timer" (555).

3) Calculate the durations of the high state and low state. Then, deduce the period T of the output signal v_s.

4) Signal v_s is applied at the input of the circuit shown in Figure E13.2. Find the evolution in time of voltages $v_S(t)$, $v_{S1}(t)$ and $v_{S2}(t)$ if R_2C_2 is very small compared to the period T.

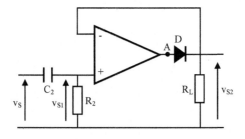

Figure E13.2.

EXERCISE 14[2]

Let us consider the circuit schematically shown in Figure E14.1.

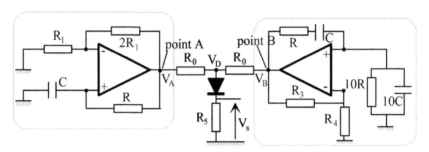

Figure E14.1.

1) Express the pulse repetition frequency of signals at points A and B. What conclusion can be drawn?

2) Draw the evolution in time of signals V_A, V_B and V_D and also that of the output signal.

3) Briefly explain the function realized by this circuit.

EXERCISE 15[3]

The task is to simultaneously generate a square signal and a triangular signal. For this purpose, the circuit shown in Figure E15.1 is the focus of interest.

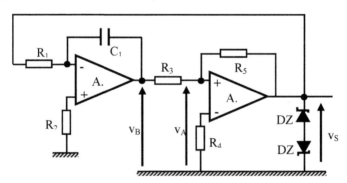

Figure E15.1.

---

2 No solution has been provided for this exercise in order to allow the reader to test themselves.
3 No solution has been provided for this exercise in order to allow the reader to test themselves.

Zener diodes D_{Z1} and D_{Z2} are identical. Their Zener voltage $V_z = 5$ V. Furthermore, $R_3 = R_5$.

1) Determine the evolution in time of v_s, v_B and v_A.

2) Calculate the periods of the square and triangular signals generated.

EXERCISE 16[4]

Let us consider the circuit schematically shown in Figure E16.1.

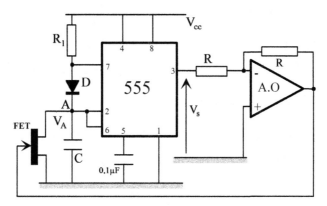

Figure E16.1.

1) Find the evolutions in time of signals v_s and v_A.

2) Determine the charge and discharge constants of capacitor C.

3) Calculate the period of the output signal v_s.

Given: $R_{DON} = R_2$ and $R_{DOFF} \rightarrow \infty$. R_{DON} and R_{DOFF} are the conduction resistance and the blocking resistance, respectively, of the field effect transistor (FET). The pinch off voltage V_p of the FET is approximately -8 V.

6.9. Solutions to exercises

SOLUTION TO EXERCISE 1

For a color version of the figures in this solution, see www.iste.co.uk/haraoubia/nonlinear1.zip.

---

4 No solution has been provided for this exercise in order to allow the reader to test themselves.

1) Expressions of voltages v_A and v_B as a function of time and time constants in the following cases:

1.1) $C_1 = 1\ \mu F$ and $C_2 = 1\ nF$

$$v_A \cong E(1-e^{\frac{-t}{RC_2}}); \qquad v_B \cong V_A = E(1-e^{\frac{-t}{RC_2}}); \text{ time constant } \tau = RC_2$$

1.2) $C_1 = 1\ nF$ and $C_2 = 1\ \mu F$

$$V_A = E(1-e^{\frac{-t}{RC_1}}); \qquad V_B = 0; \text{ time constant } \tau = RC_1$$

1.3) $C_1 = 1\ \mu F$ and $C_2 = 1\ \mu F$

$$V_A = E(1-e^{\frac{-2t}{RC_1}}); \qquad V_B = \frac{V_A}{2} = \frac{E}{2}(1-e^{\frac{-2t}{RC_1}}); \text{ time constant } \tau = RC_1/2 = RC_2/2$$

2) Evolutions of v_A and v_B as a function of time when the input voltage "v_e" is continuous and its amplitude is 5 V.

2.1) $C_1 = 1\ \mu F$ and $C_2 = 1\ nF$;

Capacitors C_1 and C_2 are connected in series. The capacitance of C_1 is 1,000 times larger than that of C_2.

Consequently, the potential drop across capacitor C_1 can be expected to be practically negligible compared to the voltage across capacitor C_2.

The two voltages v_A and v_B are practically zero.

$$v_A = (Z_{C1} + Z_{C2}).i$$
$$v_B = (Z_{C2}).i$$

$$Z_{C1} \ll Z_{C2} \Rightarrow v_A \cong v_B$$

$$v_A = E(1-e^{\frac{-t}{RC_2}})$$

$$v_B = v_A = E(1-e^{\frac{-t}{RC_2}})$$

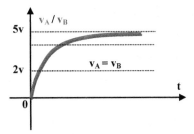

2.2) $C_1 = 1$ nF and $C_2 = 1$ μF

$$v_A = E(1 - e^{\frac{-t}{RC_1}})$$
$$v_B = 0$$

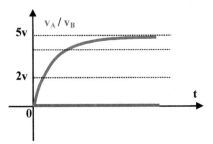

2.3) $C_1 = 1$ μF and $C_2 = 1$ μF

$$v_A = E(1 - e^{\frac{-2t}{RC_1}})$$
$$v_B = \frac{v_A}{2} = E(1 - e^{\frac{-2t}{RC_1}})$$

Time constant $\tau = RC_1/2$

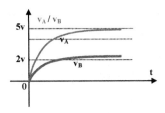

3) v_e is a unidirectional square signal. Representation of v_A and v_B and determination of the time constant.

3.1) $C_1 = 1\ \mu F$ and $C_2 = 1\ nF$

Time constant $\tau = RC_2$

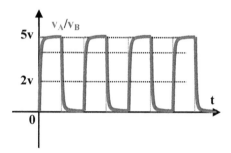

The value of the half-period is equal to $(T/2) = 0.5$ ms. The time constant is:

$\tau = RC_2 = 1\ \mu s$

Therefore, capacitor C_2 has the time to fully charge and discharge.

3.2) $C_1 = 1\ nF$ and $C_2 = 1\ \mu F$

Time constant $\tau = RC_1$

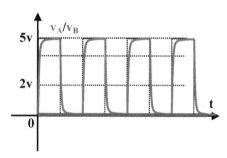

The value of the half-period is $(T/2) = 0.5$ ms. The time constant is: $\tau = RC_1 = 1\ \mu s$. Consequently, capacitor C_1 has the time to fully charge and discharge.

3.3) $C_1 = C_2 = C = 1\ \mu F$

Time constant $\tau = RC/2$

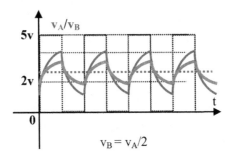

$$v_B = v_A/2$$

The value of the half-period is $(T/2) = 0.5$ ms. The time constant is:

$\tau = RC/2 = 0.5$ ms

Consequently, the equivalent capacitor $(C/2)$ does not have the time to charge or discharge.

SOLUTION TO EXERCISE 2

1) Expressions and representation of v_s and v_B.

1.1) $R_1 = 10\ \Omega$; $R_2 = 100\ k\Omega$ and $C = 1\ \mu F$

1.1.1) Expression

Resistances R_1 and R_2 are connected in series (Figure E2.4). The same current flows through each of them. Consequently, the following can be written:

$$v_S = v_B \frac{R_2}{R_1 + R_2}$$

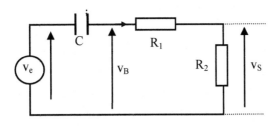

Figure E2.4.

$R_2 \gg R_1$:

$v_S = v_B$

Capacitor C charges across resistances R_1 and R_2, connected in series. Since resistance R_2 is very high compared to R_1, the effect of R_1 can be neglected.

The charge constant τ is equal to: $\tau = (R_1 + R_2).C \cong R_2.C$

The capacitor charge follows an exponential law. Initially, the capacitor is assumed to be fully discharged: $v_C(t = 0) = 0$.

$$v_C = Ae^{\frac{-t}{(R_1+R_2)C}} + B$$

$$v_C(t=0) = 0 = A + B \Rightarrow A = -B; \quad v_C(t \rightarrow \infty) = E = B$$

$$v_C = E(1 - e^{\frac{-t}{(R_1+R_2)C}}) \cong E(1 - e^{\frac{-t}{R_2C}})$$

Hence, voltages v_B and v_S can be determined.

$$v_B = v_e - v_C = Ee^{\frac{-t}{(R_1+R_2)C}} \cong Ee^{\frac{-t}{R_2C}}$$

$$v_S = v_B = Ee^{\frac{-t}{R_2C}}$$

1.1.2) Representation

$R_1 = 10\ \Omega$; $R_2 = 100\ k\Omega$ and $C = 1\ \mu F$;

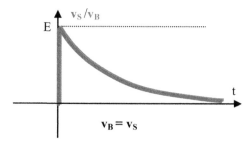

1.2) $R_1 = 100$ kΩ; $R_2 = 10$ Ω and $C = 1$ μF.

1.2.1) Expression

Resistances R_1 and R_2 are connected in series. The relation between the output voltage v_S and voltage v_B is expressed as:

$$v_S = v_B \frac{R_2}{R_1 + R_2}$$

It should be noted that resistance R_2 is very small compared to R_1. Hence:

$$v_S = v_B \frac{R_2}{R_1} \cong 0$$

Capacitor C practically charges through resistance R_1. The charge constant is equal to: $\tau = (R_1 + R_2).C \cong R_1.C$

The capacitor charge follows an exponential law starting from an initial zero voltage: $v_C(t = 0) = 0$.

$$v_C = Ae^{\frac{-t}{(R_1 + R_2)C}} + B$$

$$v_C(t = 0) = 0 = A + B \Rightarrow A = -B; \quad v_C(t \to \infty) = E = B$$

Hence, the expression of voltage across capacitor C is determined.

$$v_C = E(1 - e^{\frac{-t}{(R_1 + R_2)C}}) \cong E(1 - e^{\frac{-t}{R_1 C}}) ; \quad v_B = v_e - v_C = Ee^{\frac{-t}{(R_1 + R_2)C}} \cong Ee^{\frac{-t}{R_1 C}}$$

$$v_S = 0$$

1.2.2) Representation

$R_1 = 100\ k\Omega;\ R_2 = 10\ \Omega$ and $C = 1\ \mu F;$

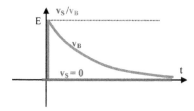

For a color version of this figure, see www.iste.co.uk/haraoubia/nonlinear1.zip

1.3) $R_1 = R_2 = R = 100\ k\Omega$ and $C = 1\ \mu F$

1.3.1) Expression

Resistances R_1 and R_2 are connected in series and have equal values. The relation between the output voltage v_S and voltage v_B is expressed as:

$$v_S = v_B \frac{R_2}{R_1 + R_2} = \frac{v_B}{2}$$

Capacitor C practically charges through resistances $R_1 + R_2$. The charge constant τ is: $\tau = (R_1 + R_2).C = 2R.C$.

The capacitor charge follows an exponential law. The initial voltage is zero: $v_C(t = 0) = 0$.

$$v_C = Ae^{\frac{-t}{(R_1 + R_2)C}} + B$$

$$v_C(t = 0) = 0 = A + B \Rightarrow A = -B;\quad v_C(t \rightarrow \infty) = E = B.$$

The following relations are obtained for the expressions of voltages across the capacitor and at point B:

$$v_C = E(1 - e^{\frac{-t}{(R_1 + R_2)C}}) = E(1 - e^{\frac{-t}{2RC}})$$

$$v_B = v_e - v_C = Ee^{\frac{-t}{(R_1 + R_2)C}} = Ee^{\frac{-t}{2RC}}$$

The output voltage is defined as:

$$v_S = \frac{v_B}{2} = \frac{E}{2} e^{\frac{-t}{2RC}}$$

1.3.2) Representation

$R_1 = 100 \text{ k}\Omega$; $R_2 = 100 \text{ k}\Omega$ and $C = 1 \text{ μF}$;

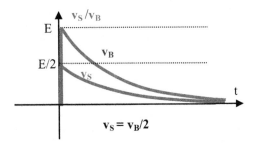

For a color version of this figure, see www.iste.co.uk/haraoubia/nonlinear1.zip

2) The input signal is a unidirectional square signal of frequency $f = 10$ kHz and amplitude E. Expressions of voltages v_s and v_B for an established steady state.

2.1) $R_1 = 10 \text{ }\Omega$; $R_2 = 100 \text{ k}\Omega$ and $C = 1 \text{ μF}$

The frequency of the input signal is 10 kHz, and the period is about 0.10 ms. The capacitor C charges through the series connection of two resistances R_1 and R_2 with the time constant: $\tau = (R_1 + R_2).C \cong 100$ ms.

In steady state, the capacitor blocks the passage of the continuous component and allows the passage toward the output of the square signal without its continuous component.

Hence, the expressions of signals v_B and v_s are determined.

$$v_S = v_B \frac{R_2}{R_1 + R_2}$$

$R_2 \gg R_1 \Rightarrow v_S = v_B$

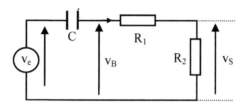

$v_C = (E/2)$ is the (continuous) average value of the input signal.

$$v_S = v_B = v_e - v_C = v_e - \frac{E}{2} = \begin{cases} \dfrac{E}{2} & 0 < t < \dfrac{T}{2} \\[2mm] \dfrac{-E}{2} & \dfrac{T}{2} < t < T \end{cases}$$

2.2) $R_1 = 100 \text{ k}\Omega$; $R_2 = 10 \ \Omega$ and $C = 1 \ \mu F$

The capacitor still charges and discharges through the series connection of the two resistances R_1 and R_2 with time constant $\tau = (R_1 + R_2).C$. The value of the time constant is very large compared to the half-period of the input signal:

$\tau \gg (T/2)$. The capacitor is applied as the continuous component of the input signal. Hence:

$$v_C = \frac{E}{2}, \qquad v_S = v_B \frac{R_2}{R_1 + R_2} \cong 0 \qquad R_2 \ll R_1$$

The expression of voltage v_B is determined similarly to the previous question.

$$v_B = v_e - v_C = v_e - \frac{E}{2} = \begin{cases} \dfrac{E}{2} & 0 < t < \dfrac{T}{2} \\[2mm] \dfrac{-E}{2} & \dfrac{T}{2} < t < T \end{cases}$$

2.3) $R_1 = R_2 = R = 100 \ \Omega$ and $C = 0.1 \ \mu F$.

The charge and discharge constant of capacitor C is equal to:

$$\tau = (R_1 + R_2).C = 2.R.C = 0.02 \text{ ms}$$

It should be recalled that the half-period of the input signal is 0.05 ms. Therefore, the capacitor charges and discharges partially.

$$v_S = v_B \frac{R_2}{R_1 + R_2} = \frac{v_B}{2} \quad \text{and} \quad v_B = v_e - v_C$$

In order to find the expression of v_B, the expressions of v_C relative to the charge and discharge of the capacitor should be found.

2.3.1) Charge case

$$v_C = Ae^{\frac{-t}{(R_1 + R_2)C}} + B$$

$$v_C(t = 0) = V_{inc} = A + B; \quad v_C(t \to \infty) = E = B \Rightarrow A = V_{inc} - E$$

where V_{inc} is the initial charge voltage (the capacitor has no time to fully charge or fully discharge).

In steady state, the voltage across the capacitor has an initial charge voltage (V_{inc}) and an initial discharge voltage (V_{ind}).

It should be noted that the final charge voltage is equal to the initial discharge voltage (V_{ind}), and the final discharge voltage is equal to the initial charge voltage (V_{inc}):

$$v_C = (V_{inc} - E)e^{\frac{-t}{(R_1 + R_2)C}} + E = (V_{inc} - E)e^{\frac{-t}{2RC}} + E$$

When the input voltage is at high state, the expression of v_B is:

$$v_B = v_e - v_C$$

$$v_B = (v_e - v_C) = (E - V_{inc})e^{\frac{-t}{2RC}}$$

2.3.2) Discharge case

$$v_C = A_1 e^{\frac{-t}{(R_1 + R_2)C}} + B_1$$

The boundary conditions allow the determination of coefficients A and B.

$$v_C(t=0) = V_{ind} = A_1 + B_1; \quad v_C(t \to \infty) = 0 = B \Rightarrow A = V_{ind} ;$$

where V_{ind} is the initial discharge voltage.

Finally, the voltage across capacitor C is defined by the following relation:

$$v_C = V_{ind}e^{\frac{-t}{(R_1+R_2)C}} = V_{ind}e^{\frac{-t}{2RC}}$$

When the input voltage is in low state (capacitor discharge), the voltage v_B has the following expression:

$$v_B = v_e - v_C = -v_C = -V_{ind}e^{\frac{-t}{2RC}}$$

2.3.3) Determination of V_{inc} and V_{ind}

The initial charge voltage V_{inc} is also the final discharge voltage:

$$V_{inc} = V_{ind}e^{\frac{-T/2}{2RC}} = 0.082V_{ind}$$

The initial discharge voltage V_{ind} is also the final charge voltage:

$$V_{ind} = (V_{inc} - E)e^{\frac{-T/2}{2RC}} + E = 0.082V_{inc} + 0.918E$$

$$V_{ind} = 0.0067V_{ind} + 0.918E \Rightarrow V_{ind} \cong 0.924E$$

$$V_{inc} \cong 0.076E$$

The expression of v_B and v_S can then be fully determined. Indeed, when $v_e = E$:

$$v_B = (v_e - v_C) = 0.924.E.e^{\frac{-t}{2RC}} \quad \text{and} \quad v_S = \frac{v_B}{2}$$

When $v_e = 0$:

$$v_B = -V_{ind}e^{\frac{-t}{2RC}} = -0.924.E.e^{\frac{-t}{2RC}} \quad \text{and} \quad v_S = \frac{v_B}{2}$$

3) Representation as a function of time of signals v_e, v_s and v_B for the three cases mentioned in question 2.

3.1) $R_1 = 10\ \Omega$, $R_2 = 100\ k\Omega$ and $C = 1\ \mu F$,

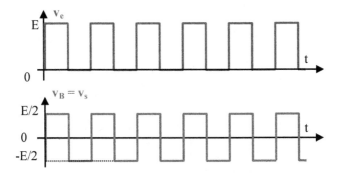

For a color version of this figure, see www.iste.co.uk/haraoubia/nonlinear1.zip

3.2) $R_1 = 100\ k\Omega$, $R_2 = 10\ \Omega$ and $C = 1\ \mu F$

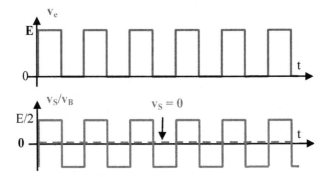

For a color version of this figure, see www.iste.co.uk/haraoubia/nonlinear1.zip

3.3) $R_1 = R_2 = R = 100\ \Omega$ and $C = 0.1\ \mu F$.

Let us recall that the calculations effected have allowed the definition of the expressions of various voltages (v_B and v_S) as a function of time, both during the charge and the discharge of the capacitor, leading to the following relations:

During the charge:

$$v_B = 0.924 E e^{\frac{-t}{2RC}} \quad \text{and} \quad v_S = \frac{v_B}{2}$$

During the discharge:

$$v_B = -V_{ind} e^{\frac{-t}{2RC}} = -0.924 E e^{\frac{-t}{2RC}} \text{ and } v_S = \frac{v_B}{2}$$

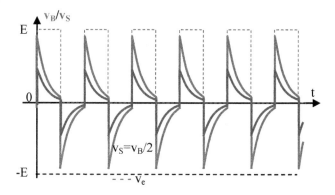

For a color version of this figure, see www.iste.co.uk/haraoubia/nonlinear1.zip

SOLUTION TO EXERCISE 3

1) Representation of the evolution in time of the signals across resistances R_1 and R_2 and across capacitor C when the time constant $\tau_1 = R_2C$ is very large compared to period T.

Writing the voltages in the circuit algebraically yields:

$$v_e = v_{R1} + v_{R2} + v_C$$

Given that the time constant is very large compared to the period of the input signal, when the steady state is established, the average value of the input signal is found across capacitor C.

This average value is defined as follows:

$$v_C = \frac{1}{T}\int_0^T v_e \, dt = \frac{1}{T}\int_0^{T/2} V \, dt = \frac{V}{2}$$

The determination of voltage across C allows the deduction of the variations of voltage across R_1 and the voltage across R_2:

$$v_C = (V/2); \quad v_e - v_C = v_{R1} + v_{R2} = v_e - (V/2); \quad \frac{v_{R1}}{v_{R2}} = \frac{R_1}{R_2}$$

If, for the sake of simplicity, a choice is made for $R_1 = R_2$, then:

$$v_{R1} = v_{R2} = \frac{v_e - \dfrac{V}{2}}{2}$$

The various relations established can be used to draw the graphic representations of v_{R1}, v_{R2} and v_C ($R_1 = R_2$), as schematically shown below.

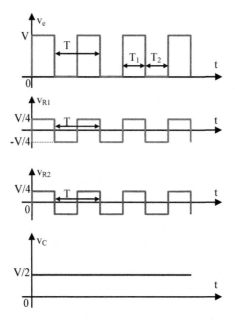

For a color version of this figure, see www.iste.co.uk/haraoubia/nonlinear1.zip

2) Evolution of the signals across resistances R_1 and R_2 and capacitor C. Function of the circuit when the output is considered across resistance R_2.

The time constant is very low compared to the period of the input signal and the capacitor has the time to fully charge and discharge.

$$v_C = \begin{cases} V(1 - e^{\frac{-t}{\tau}}) & \text{charge} \\ Ve^{\frac{-t}{\tau}} & \text{discharge} \end{cases} \qquad \text{with } \tau = (R_1 + R_2)C$$

To simplify, let us consider $R_1 = R_2$: $v_e - v_C = v_{R1} + v_{R2} = 2v_{R1}$

$$v_{R1} = \begin{cases} \dfrac{V}{2}e^{\frac{-t}{\tau}} & \text{charge} \\ \dfrac{-V}{2}e^{\frac{-t}{\tau}} & \text{discharge} \end{cases}$$

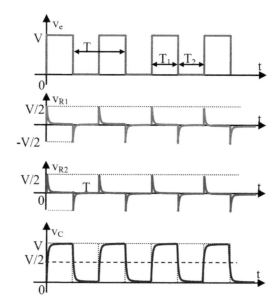

For a color version of this figure, see www.iste.co.uk/haraoubia/nonlinear1.zip

When the output is taken across R_2, it can be said that the function of the circuit is a high-pass filter.

SOLUTION TO EXERCISE 4

1) Charge and discharge constants. Comparison with the half-period of the input signal.

1) a) $R_1 = R_2 = 1 \text{ k}\Omega$, $C = 25 \text{ nF}$

The capacitor charges and discharges through the same path.

$$\tau_c = \tau_d = (R_1 + R_2)C = 50 \ \mu s$$

The period of the input signal is given by:

$$T = (1/F) = 1 \text{ ms}; \ T/2 = 500 \ \mu s \Rightarrow \tau_c = \tau_d \ll (T/2).$$

The capacitor has the time to fully charge and discharge during a half-period.

1) b) $R_1 = R_2 = 10 \text{ k}\Omega$, $C = 20 \ \mu F$;

$$\tau_c = \tau_d = (R_1 + R_2)C = 40 \text{ ms and } T/2 = 500 \ \mu s; \ \tau_c = \tau_d \gg (T/2).$$

The capacitor does not have the time to fully charge and discharge during a half-period.

2) Expressions of voltages v_{R1}, v_A and v_S and of the voltage across capacitor C, and their representation when $R_1 = R_2 = 1 \text{ k}\Omega$ and $C = 25 \text{ nF}$.

2.1) During charge ($v_e = E$)

The capacitor has the time to charge

$$v_C = E(1 - e^{-\frac{t}{(R_1 + R_2)C}}) \text{ and } v_A = v_C + v_S; \ v_{R1} = v_e - v_A = v_e - (v_C + v_S)$$

$$v_S = v_{R2} \text{ and } v_S + v_{R1} = v_e - v_C; \ v_{R2} = v_{R1} = v_S \text{ et } 2v_S = v_e - v_C \ (R_1 = R_2)$$

Finally, the expressions of voltages v_{R1} and v_S can be defined as follows:

$$v_{R1} = \frac{E}{2} e^{-\frac{t}{(R_1 + R_2)C}}); \quad v_S = \frac{E}{2} e^{-\frac{t}{(R_1 + R_2)C}}); \quad v_A = E(1 - \frac{1}{2} e^{-\frac{t}{(R_1 + R_2)C}})$$

2.2) During the discharge ($v_e = 0$)

The capacitor has the time to discharge

$$v_C = E.e^{-\frac{t}{(R_1+R_2)C}} \; ; \; \quad v_{R1} = -\frac{E}{2}.e^{-\frac{t}{(R_1+R_2)C}}$$

$$v_S = -\frac{E}{2}e^{-\frac{t}{(R_1+R_2)C}} \; ;$$

$$v_A = v_S + v_C = \frac{E}{2}e^{-\frac{t}{(R_1+R_2)C}}$$

Representation of voltages v_e, v_{R1}, v_A, v_S and v_C as a function of time

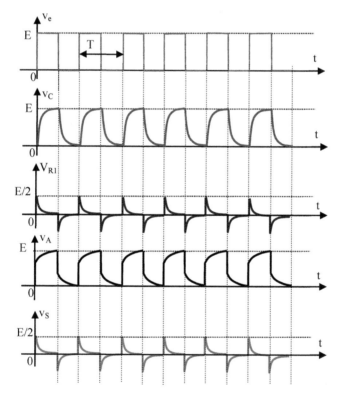

3) Expressions of voltages v_{R1}, v_A and v_S and of the voltage across capacitor C, and their representation in case $R_1 = R_2 = 10$ kΩ and $C = 20$ μF.

In steady state, the continuous component (average value) of the input signal is found across capacitor C.

$$v_C = \frac{1}{T}\int_0^T v_e \, dt = \frac{1}{T}\int_0^{T/2} E \, dt = \frac{E}{2} \; ;$$

$$v_C = \frac{E}{2} \; ;$$

$$v_A = v_C + v_S \; ;$$

$$v_S = v_{R2} \; ;$$

$$v_S + v_{R1} = v_e - v_C$$

$R_1 = R_2$ and R_1 and R_2 are connected in series:

$$v_{R2} = v_{R1} = v_S$$

This allows the determination of the expression of v_S and, consequently, of v_{R1}:

$$v_S = \frac{1}{2}(v_e - v_C) = \begin{cases} \dfrac{E}{4} & v_e = E \\[2mm] \dfrac{-E}{4} & v_e = 0 \end{cases}$$

$$v_{R1} = \begin{cases} \dfrac{E}{4} & v_e = E \\[2mm] \dfrac{-E}{4} & v_e = 0 \end{cases}$$

$$v_A = v_C + v_S = \begin{cases} \dfrac{3E}{4} & v_e = E \\[2mm] \dfrac{E}{4} & v_e = 0 \end{cases}$$

The diagram of various signals (v_e, v_C, v_{R1}, v_A and v_S) is shown in the following figure:

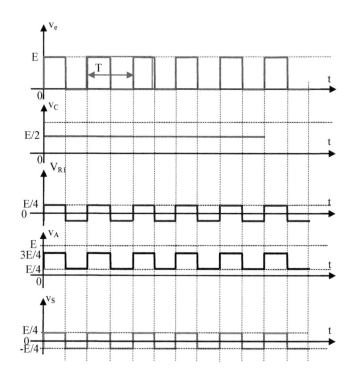

SOLUTION TO EXERCISE 5

1) Expressions of v_{C1}, v_{C2} and v_R when the voltage v_e is continuous and equal to $E = 5$ V

$$v_{C1} = Z_1 i \text{ and } v_{C2} = Z_2 i.$$

The capacitances of capacitors C_1 and C_2 are equal, and the same current flows through them. Hence:

$$Z_1 = Z_2 \text{ and } v_{C1} = v_{C2} = v_C.$$

The two capacitors are connected in series; therefore, they can be joined in only one capacitor. The studied circuit, which is schematically shown in Figure E5.2(a), has an equivalent circuit shown in Figure E5.2(b).

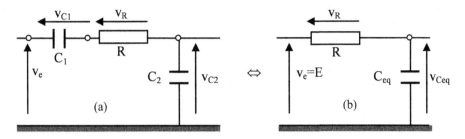

Figure E5.2.

$$C_1 = C_2, v_{C1} = v_{C2} = v_C; \quad C_{eq} = \frac{C_1 . C_2}{C_1 + C_2} = \frac{C}{2};$$

$$E = 2v_C + v_R. = v_R + v_{Ceq} \quad \text{and} \quad v_{Ceq} = Ae^{\frac{-t}{\tau}} + B \quad \text{with} \quad \tau = \frac{RC}{2}$$

$$t = 0, A + B = 0; t \rightarrow \infty, V_{Ceq} = E$$

$$v_{Ceq} = E(1 - e^{\frac{-t}{\tau}}), \quad v_{C1} = v_{C2} = \frac{v_{Ceq}}{2} = \frac{E}{2}(1 - e^{\frac{-t}{\tau}}), \quad v_R = E - v_{Ceq} = Ee^{\frac{-t}{\tau}}$$

2) Representation of voltages v_{C1}, v_{C2} and v_R as a function of time (see Figure E5.3) and final values ($t \rightarrow \infty$) of these voltages.

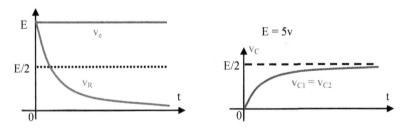

Figure E5.3. *For a color version of this figure, see*
www.iste.co.uk/haraoubia/nonlinear1.zip

The final value of v_{C1} = final value of v_{C2} = 2.5 V; final value of v_R = 0 V

3) Expressions and representation as a function of time of v_{C1}, v_R and v_{C2}, knowing that v_e is a square voltage of amplitude E = 5 V and frequency equal to 100 Hz.

The charge and discharge time constants are equal to τ: $\tau = \dfrac{RC}{2} = 0.5$ ms. The period of the input signal is: $T = 10$ ms.

Consequently, capacitors C_1 and C_2 have the time to fully charge and discharge. Since their series connection has a capacitance that is half of C_1 or C_2, the equivalent capacitance C_{eq} also has the time to fully charge and discharge.

3.1) Expressions of v_{C1}, v_R and v_{C2}

During the charge:

$$v_{Ceq} = E(1 - e^{\frac{-t}{\tau}}); \quad v_{C1} = v_{C2} = \frac{v_{Ceq}}{2} = \frac{E}{2}(1 - e^{\frac{-t}{\tau}}); \quad v_R = Ee^{\frac{-t}{\tau}}$$

During the discharge:

$$v_{Ceq} = Ee^{\frac{-t}{\tau}}); \quad v_{C1} = v_{C2} = \frac{v_{Ceq}}{2} = \frac{E}{2}e^{\frac{-t}{\tau}}; \quad v_R = -Ee^{\frac{-t}{\tau}}$$

3.2) Representation of v_e, v_{C1}, v_{C2} and v_R (see Figure E5.4)

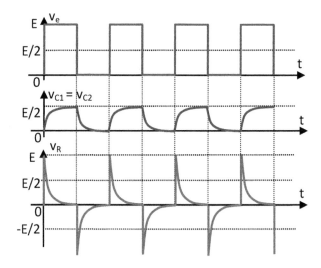

Figure E5.4. *For a color version of this figure, see*
www.iste.co.uk/haraoubia/nonlinear1.zip

SOLUTION TO EXERCISE 6

1) Expressions of v_S and v_A

$$\text{At } t = 0, \; v_C = 0; \; v_S = \frac{R}{R+R}v_e = \frac{E}{2}, \; v_A = v_S = \frac{E}{2}$$

When time t evolves, the various expressions can be written as follows:

$$v_C = E(1-e^{-\frac{t}{2RC}}) \; ; v_S = \frac{E}{2}e^{-\frac{t}{2RC}}) \text{ and}$$

$$v_A = v_C + v_S = E(1-\frac{1}{2}e^{-\frac{t}{2RC}})$$

2) Representations of v_S and v_A (see Figure E6.2)

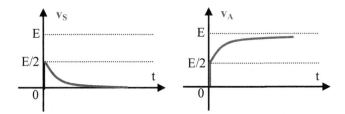

Figure E6.2. *For a color version of this figure, see www.iste.co.uk/haraoubia/nonlinear1.zip*

3) Expressions of the charge and discharge constants of v_S and v_A for the case when: $R = 1 \text{ k}\Omega$ and $C = 0.1 \text{ }\mu\text{F}$. v_e is a square signal.

The calculation of the charge and discharge constants yields:

$$\tau_{charge} = \tau_{discharge} = \tau = 2.RC = 0.2 \text{ ms}; \; T = 10 \text{ ms} \Rightarrow T >> \tau)$$

Capacitor C has the time to fully charge and discharge.

3.1) Case of the charge of capacitor C

$$v_{C(ch\,arg\,e)} = E(1-e^{-\frac{t}{2RC}})$$

The evolutions of output voltages and of the voltage at point A during the charge of capacitor C are expressed by the following relations:

$$v_{S(charge)} = \frac{E}{2}e^{-\frac{t}{2RC}}, \qquad v_{A(charge)} = v_C + v_S = E(1 - \frac{1}{2}e^{-\frac{t}{2RC}})$$

3.2) Case of discharge of capacitor C

$$v_{C(discharge)} = Ee^{-\frac{t}{2RC}}$$

$$v_{S(discharge)} = -\frac{E}{2}e^{-\frac{t}{2RC}}), \qquad v_{A(discharge)} = v_C + v_S = \frac{E}{2}e^{-\frac{t}{2RC}})$$

4) Representation of the evolution in time of v_S and v_A (see Figure E6.3).

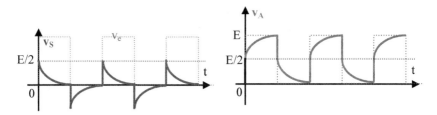

Figure E6.3. *For a color version of this figure, see www.iste.co.uk/haraoubia/nonlinear1.zip*

5) Expressions of the charge and discharge constants of v_S and v_A and representation of their evolution when R = 100 kΩ and C = 0.5 μF, and the function of the circuit.

5.1) Charge and discharge time constants

$$\tau_{charge} = \tau_{discharge} = \tau = 2\,RC = 100 \text{ ms}, \ T = 10 \text{ ms (T period) } T \gg \tau.$$

5.2) Expressions of v_S and v_A

5.2.1) Case when $v_e = E$

$$v_C = E/2; \quad v_s = E/4$$

$$v_A = v_C + v_S = 3E/4$$

5.2.2) Case when $v_e = 0$

$$v_C = E/2; \quad v_S = -E/4; \quad v_A = v_C + v_S = E/4$$

5.3) Representations of v_S and v_A (see Figure E6.4)

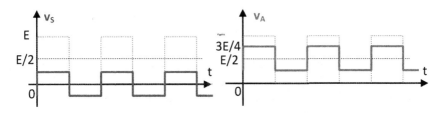

Figure E6.4. *For a color version of this figure, see*
www.iste.co.uk/haraoubia/nonlinear1.zip

5.4) Function of the circuit: the circuit studied in question 5 is a coupling circuit.

SOLUTION TO EXERCISE 7

1) Expressions and representations as a function of time of signals v_{R1} and v_s

1.1) Expressions of signals v_{R1} and v_s

The diode is assumed ideal. The negative part of the input voltage is clipped. Therefore, only the positive part remains.

Signals v_{R1} and v_S are periodic

$$v_{R1} = \begin{cases} E & 0 < t < T_1 \\ 0 & T_1 < t < T \end{cases}$$

$$v_S = \begin{cases} E + (V_{ic} - E)e^{\frac{-t}{\tau_1}} & 0 < t < T_1 \quad \text{charge} \\ V_{id}.e^{\frac{-t}{\tau_1}} & T_1 < t < T \quad \text{discharge} \end{cases}$$

where V_{ic} is the initial charge voltage and V_{id} is the initial discharge voltage.

In order to find the charge constant (τ_1) and the discharge constant (τ_2), it is sufficient to note that the capacitor charges across resistance R_2 and dynamic resistance R_d of the diode (here $R_d = 0$). On the contrary, it discharges through the series connection of resistances R_1 and R_2. Hence:

$$\tau_1 = R_2.C \text{ and } \tau_2 = (R_1 + R_2).C$$

1.2) Representations of v_{R1} and v_S (see Figure E7.3)

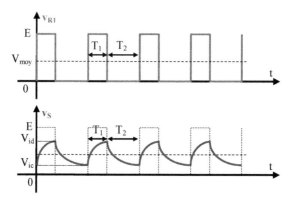

Figure E7.3.

2) Representation of $v_{R2}(t)$ (see Figure E7.4)

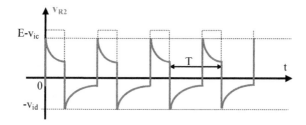

Figure E7.4.

It can be noted without difficulty that $\mathbf{v_{R2}} = \mathbf{v_{R1}} - \mathbf{v_S}$

$$v_{R2} = \begin{cases} (E - V_{ic})e^{\frac{-t}{\tau_1}} & 0 < t < T_1 \quad \text{charge} \\ -V_{id}.e^{\frac{-t}{\tau_2}} & T_1 < t < T \quad \text{discharge} \end{cases}$$

SOLUTION TO EXERCISE 8

1) Expressions of voltages at points v_{e+}, v_C and v_S.

The output of the operational amplifier is assumed at high state, the diode being considered ideal. Under these conditions: $v_S = V_{CC}$

The expression of the voltage applied at the non-inverting input of the operational amplifier is:

$$v_{e+} = \frac{R_1}{R_1 + R_2} v_S \quad v_{e+} = \frac{R_1}{R_1 + R_2} V_{CC}$$

The capacitor starts to charge from a voltage that is initially zero.

$$v_C = A e^{\frac{-t}{R3C}} + B$$

where A and B are constants that can be determined using boundary conditions. At $t = 0$, $v_C = 0$ and when $t \to \infty$, $v_C \to V_{CC}$; $A + B = 0$ and $B = V_{CC}$; $A = -V_{CC}$

$$v_C = V_{CC}(1 - e^{\frac{-t}{R3C}})$$

2) Value of voltage v_{e-} starting from which the output voltage switches. Expressions of v_{e+}, v_C and v_S right after the switching and their value and representation as a function of time.

2.1) Value of voltage v_{e-} starting from which the output voltage shifts

The voltage across capacitor increases and is compared at any moment with the voltage on the non-inverting input of the operational amplifier. The output switches as soon as the voltage v_{e-} reaches and slightly exceeds voltage v_{e+}. The expression of the switching voltage is:

$$v_{e-} = v_C = \frac{R_1}{R_1 + R_2} v_S = \frac{V_{CC}}{2}$$

2.2) Expressions of v_{e+}, v_C and v_S right after the switching and their value.

When there is a switching, the output of the operational amplifier passes from high saturation to low saturation. The presence of diode D (ideal diode) clips the signal and hence at the output: $v_S = 0$

This yields: $v_{e+} = 0$

The output voltage v_S ($v_S = 0$) is applied through resistance R_3 to the capacitor previously charged under voltage ($V_{CC}/2$). This capacitor starts its discharge seeking to reach zero value. Since the voltage across C cannot decrease below zero, switching is no longer possible. The evolution of voltage across the capacitor can be described as follows:

$$v_C = A_1 e^{\frac{-t}{R3C}} + B_1 \; ;$$

for $t = 0$, $v_C = \dfrac{V_{CC}}{2}$ and for $t \to \infty$ $v_C = 0$, $B = 0$ and $A = \dfrac{V_{CC}}{2}$

$$v_C = \frac{V_{CC}}{2} e^{\frac{-t}{R3C}} \; ;$$

2.3) Representations of evolutions of voltages v_{e+}, v_C and v_S (see Figure E8.2)

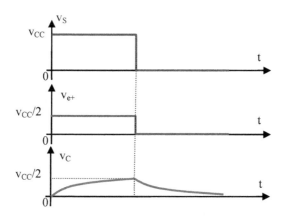

Figure E8.2.

3) Expressions of various voltages (v_{e+}, v_C and v_S) and representations of their evolution in time knowing that the diode threshold voltage is $V_0 = 0.6$ V. The output of the amplifier is assumed at high state:

$$v_S = V_{CC}, \quad v_{e+} = \frac{R_1}{R_1 + R_2} V_{CC} = \frac{V_{CC}}{2} \; ; \quad v_C = V_{CC}(1 - e^{\frac{-t}{R3C}})$$

As soon as v_C reaches $(V_{CC}/2)$, switching occurs. Hence:

$$v_S = -V_0, \; v_{e+} = -(V_0/2).$$

The capacitor starts to discharge, and the voltage across it decreases from a value of $(V_{CC}/2)$ seeking to reach $-V_0$ according to the law:

$$v_C = A_2 e^{\frac{-t}{R_3 C}} + B_2 \, ;$$

for $t = 0$, $v_C = \dfrac{V_{CC}}{2}$ and for $t \to \infty$, $v_C = -V_0$, $B_2 = -V_0$ and $A_2 = \dfrac{V_{CC}}{2} + V_0$.

Finally, the voltage across the capacitor during discharge can be written as follows:

$$v_C = (\frac{V_{CC}}{2} + V_0) e^{\frac{-t}{R_3 C}} - V_0$$

The voltage across the capacitor never reaches the value $-V_0$, since at the moment when it reaches $(-V_0/2)$ by a lower value, there is a switching of the output of the operational amplifier from the low state to the high state: $v_S = V_{CC}$.

From this instant, the capacitor starts its charge from an initial voltage of $(-V_0/2)$ seeking to reach $(+V_{CC})$ according to the law:

$$v_C = A_3 e^{\frac{-t}{R_3 C}} + B_3 \, ;$$

for $t = 0$, $v_C = \dfrac{-V_0}{2} = A_3 + B_3$ and for $t \to \infty$, $v_C = V_{CC}$, $B_3 = V_{CC}$ and

$$A_3 = \frac{-V_0}{2} - V_{CC}$$

$$v_C = (\frac{-V_0}{2} - V_{CC}) e^{\frac{-t}{R_3 C}} + V_{CC}$$

The voltage across capacitor ($v_C = v_{e-}$) increases until it reaches $(V_{CC}/2)$ by higher value.

At this instant, there is a switching of the output from the high state to the low state:

$$v_S = -V_0 \text{ and } v_{e+} = -(V_0/2).$$

The cycle thus described repeats infinitely.

The representation of signals present at the output, across the capacitor and at the non-inverting input of the operational amplifier is given by the diagram shown in Figure E8.3.

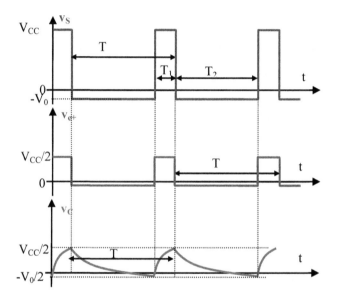

Figure E8.3. *Representation of various signals: v_S, v_{e^+}, v_{e^-} (v_C)*

4) Type of signal obtained at the output and signal frequency

4.1) Type of signal

The output signal is rectangular. It has a non-zero average value. This signal evolves between two extreme values (V_{CC} and $-V_0$).

The rectangular shape is due to the presence of the diode causing the switching thresholds to be non-symmetrical. The switching thresholds are ($V_{CC}/2$) and ($-V_0/2$).

4.2) Determination of the frequency of the output signal

In order to obtain the frequency **F** of the output signal, it is sufficient to calculate the period **T** of this signal.

$$T = T_1 + T_2 \text{ and } F = (1/T)$$

For the calculation of T_1, it is sufficient to note that this duration is the time needed by the capacitor C to charge from an initial value of $-V_0/2$ up to a final value of $(V_{CC}/2)$. Let us recall that the law that governs the charge of the capacitor is expressed by the following relation:

$$v_C = (\frac{-V_0}{2} - V_{CC})e^{\frac{-t}{R_3C}} + V_{CC}$$

For $t = T_1$, $v_C = V_{CC}/2$.

$$v_C = (\frac{-V_0}{2} - V_{CC})e^{\frac{-T_1}{R_3C}} + V_{CC} = \frac{V_{CC}}{2}$$

$$T_1 = R_3C\text{Ln}\left[\frac{2V_{CC} + V_0}{V_{CC}}\right]$$

For the calculation of T_2, it is sufficient to see that this duration is the time needed by capacitor C to discharge from an initial value of $V_{CC}/2$ to a final value of $(-V_0/2)$.

The law that governs the capacitor discharge is expressed by the following relation:

$$v_C = (\frac{V_{CC}}{2} + V_0)e^{\frac{-t}{R_3C}} - V_0$$

For $t = T_2$, $v_C = -V_0/2$.

$$v_C(T_2) = (\frac{V_{CC}}{2} + V_0)e^{\frac{-T_2}{R_3C}} - V_0 = \frac{-V_0}{2}$$

$$T_2 = R_3C.\text{Ln}\left[\frac{V_{CC} + 2V_0}{V_0}\right]$$

According to the relations defining T_1 and T_2, it can be noted that the duration of the low state is significantly longer than the duration of the high state.

The period T of the output signal v_S is:

$$T = T_1 + T_2 = R_3 C . Ln \left[\left(\frac{2V_{CC} + V_0}{V_{CC}} \right) \left(\frac{V_{CC} + 2V_0}{V_0} \right) \right] \text{ and } F = (1/T)$$

SOLUTION TO EXERCISE 9

1) Diagram and charge and discharge paths of capacitor C (see Figure E9.3)

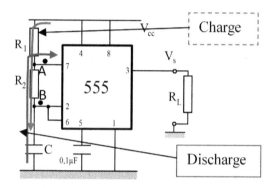

Figure E9.3. *Capacitor charges across resistances R_1 and R_2 and discharges across R_2. For a color version of this figure, see www.iste.co.uk/haraoubia/nonlinear1.zip*

2) Operation and function of the circuit

This circuit has two switching thresholds imposed by the internal diagram of 555. These switching thresholds, which are $(V_{CC}/3)$ and $(2V_{CC}/3)$, act on two comparators that in turn drive an SR flip-flop.

The state of the inputs of the SR flip-flop sets the state of the output of the studied circuit. Capacitor C charges across R_1 and R_2 when the output is at high state. The complementary output of the SR flip-flop is at low state. Transistor Tr is blocked and has a very high collector–emitter resistance. When the charge of the capacitor reaches the threshold $(2V_{CC}/3)$, the output goes from the high state to the low state. The complementary output is at the high state.

Transistor Tr is saturated and its collector is connected to zero potential. Consequently, capacitor C discharges through R_2. As soon as the voltage across the capacitor reaches the value ($V_{CC}/3$), the output shifts back from the low state to the high state.

The described cycle is repeated to allow the generation of a rectangular signal at the output. The function of the circuit: generation of rectangular waves.

3) Representations of the evolutions of signals v_A, v_B and v_S (see Figure E9.4).

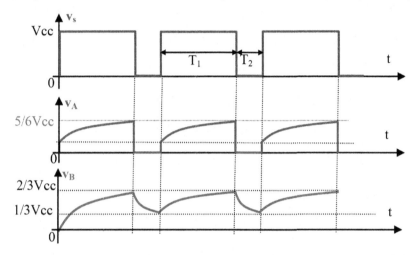

Figure E9.4. *Representations of the evolutions of signals v_A, v_B and v_S*

4) Expressions and values of the frequency of the output signal and of the cyclic ratio with $R_1 = 10$ kΩ, $R_2 = 2$ kΩ and $C = 4.7$ nF

4.1) Frequency of the output signal

$T_1 = (R_1 + R_2).C.Ln(2)$ and $T_2 = R_2.C.Ln(2)$.

$T = (R_1 + 2R_2).C.Ln(2)$

$$f = \frac{1}{(R_1 + 2R_2).C.Ln(2)}$$

4.2) Duty cycle

$$r = \frac{T_1}{T} = \frac{(R_1 + R_2)}{(R_1 + 2R_2)}$$

Numerical application: $f = 21.9\text{KHz}; \ r = 0.85$

SOLUTION TO EXERCISE 10

1) State of diode D and expressions and values of v_A and v_B at $t = 0$.

Given the condition imposed on voltage $v_S = +V_{CC} = 14$ V at instant $t = 0$, the diode D can only be blocked.

$$v_B = 0$$

$$v_A = \frac{R}{R + 6R} v_S = \frac{R}{R + 6R} V_{CC} = \frac{V_{CC}}{7} = 2V$$

2) Expression of the charge constant (τ_c) and discharge constant (τ_d) of capacitor C

The capacitor charges and discharges through the same path:

$$\tau_c = \tau_d = RC$$

3) Expressions and values of the switching thresholds

3.1) High threshold V_1

When $v_S = V_{CC}$, the diode D is blocked: $v_A = V_1 = \frac{R}{R + 6R} V_{CC}$

$$V_1 = \frac{V_{CC}}{7} = 2V$$

3.2) Low threshold V_2

When $v_S = -V_{CC}$, the diode D is conducting: $v_A = V_2 = \frac{-R}{R + R_{eq}} V_{CC}$, with

$$R_{eq} = \frac{12R.6R}{18R} = 4R, \quad V_A = V_2 = \frac{-R}{R + 4R} V_{CC} = \frac{-V_{CC}}{5} = -2.8V$$

$$V_2 = \frac{-V_{CC}}{5} = -2.8V$$

4) Representation of the evolution of v_A, v_B and v_S, and function of the circuit

4.1) Representation of $v_A(t)$, $v_B(t)$ and $v_S(t)$ (see Figure E10.2).

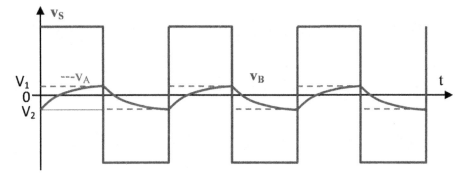

Figure E10.2. *Representation of v_A, v_B and v_S. For a color version of this figure, see www.iste.co.uk/haraoubia/nonlinear1.zip*

4.2) Function of the circuit: Astable – generator of rectangular wave.

SOLUTION TO EXERCISE 11

1) Truth table of NAND gate

A	F	$\overline{AF}$
0	0	1
0	1	1
1	0	1
1	1	0

2) Study of the circuit operation by drawing the evolutions of voltages v_B, v_D, v_F and v_S and of the voltage across capacitance C (v_C).

2.1) Stage 1

Presuming that the output of gate 2 is at the high state:

$$v_S = V_{CC} \Rightarrow V_D = 0$$

Voltage v_S was previously in the low state and then it went to the high state.

$$v_S: 0 \rightarrow +V_{CC} \text{ (variation of } \Delta v_S = +V_{CC}).$$

It should be noted that when v_S was in idle state ($v_S = 0$), $v_D = V_{CC}$ and consequently $v_B = v_F = V_{CC}$. As soon as voltage is applied at the second input of the first NAND gate:

$$v_D = 0, v_S = V_{CC} \text{ and } v_B = v_F = 2V_{CC} \text{ (transient state)}$$

Under these conditions, the diagram of the circuit shown in Figure E11.2 is applicable.

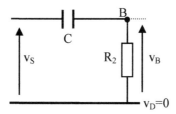

Figure E11.2.

Voltage v_B tends to decrease in order to cancel out the current in the circuit (charge of capacitor C) until v_B is slightly below $V_{CC}/3$. At this moment, voltage v_D goes from zero to V_{CC} and the output voltage v_S goes instantaneously from $+V_{CC}$ to zero ($\Delta v_S = -V_{CC}$).

This variation ($-V_{CC}$) is directly transmitted to point B through capacitor C. This yields:

$$v_D = V_{CC}, v_S = 0 \text{ and } v_B = v_F = (V_{CC}/3) - V_{CC} = (-2V_{CC}/3)$$

It is the beginning of the steady state. Voltages v_B and v_F are equal, there is no potential drop across R_1 and the current inflow at the gates is practically zero.

$$v_F = (-2V_{CC}/3) \cong \text{"0" (logical zero)}$$

2.2) Stage 2

Under the previously mentioned conditions and depending on the values of voltages existing at this stage, the circuit can be schematically drawn using Figure E11.3. Voltage v_B increases starting from the prior initial value of $(-2V_{CC}/3)$, seeking to reach V_{CC}.

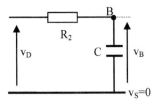

Figure E11.3.

The capacitor seeks to recover all the voltage applied to it through resistance R_2 to cancel out the current in the circuit.

Voltage v_B (and consequently voltage v_F) increases exponentially.

When voltage v_F increases and slightly exceeds $(V_{CC}/3)$, voltage v_D goes from $+V_{CC}$ to zero and voltage v_S goes from zero to $+V_{CC}$ (variation $\Delta v_S = +V_{CC}$). This variation is instantaneously transmitted to point B. Hence:

$$v_D = 0, \; v_S = V_{CC} \text{ and } v_B = v_F = (V_{CC}/3) + V_{CC} = (4V_{CC}/3)$$

2.3) Stage 3

Given the presence of these voltages, the circuit comprising the capacitor can be schematically shown in Figure E11.4.

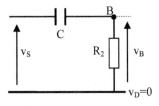

Figure E11.4.

This is the circuit obtained at stage 1. The only difference resides in the initial value of voltage $v_B = (4V_{CC}/3)$. This voltage v_B should therefore evolve until reaching $(V_{CC}/3)$ by lower value. The system switches and hence:

$$v_D = V_{CC}, v_S = 0; v_B = v_F = (V_{CC}/3) - V_{CC} = (-2V_{CC}/3)$$

Thus, the system goes indefinitely from stage 2 to stage 3, and vice versa, to allow for the generation of a rectangular signal at the output.

The various signals involved (v_S, v_B, v_F, v_C and v_D) are schematically shown in Figure E11.5.

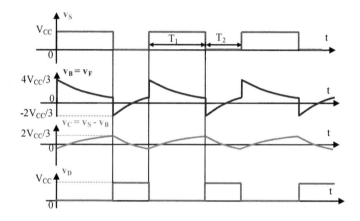

Figure E11.5. *Evolutions of voltages v_S, v_B, v_F, v_C and v_D*

3) Expressions of v_A, v_B and v_C when the steady state has been reached:

3.1) $v_S = V_{CC}$

$$v_A = V_{CC}, \quad v_B = \frac{4}{3}V_{CC} \cdot e^{\frac{-t}{R_2 C}}, \quad v_C = v_S - v_B = V_{CC}(1 - \frac{4}{3}) \cdot e^{\frac{-t}{R_2 C}}$$

3.2) $v_S = 0$

$$v_A = V_{CC}, v_B = V_{CC}\left(1 - \frac{5}{3}\right) \cdot e^{\frac{-t}{R_2 C}}, \quad v_C = v_S - v_B = -V_{CC}(1 - \frac{5}{3}) \cdot e^{\frac{-t}{R_2 C}}$$

4) Function of the circuit and its state when $R_1 = 0$.

The circuit is an astable multivibrator. When $R_1 = 0$, the NAND gate number 1 is no longer protected in terms of current; therefore, it runs the risk of being destroyed as soon as the power supply is on.

5) Period T of the output signal, expression, value of the cyclic ratio and nature of the output signal

5.1) Period of the output signal

The period can be calculated by determining the duration T_1 of the high state and the duration T_2 of the low state since:

$$T = T_1 + T_2$$

5.1.1) Calculation of the duration T_1 of the high state

The duration T_1 is the time needed for voltage v_B to go from $(4V_{CC}/3)$ to $(V_{CC}/3)$. Voltage v_B evolves according to an exponential law (that follows the charge or discharge of a capacitor). The charge or discharge constant for this circuit is $\tau = R_2.C$.

The evolution of v_B can then be described by the following law:

$$v_B = A.e^{\frac{-t}{\tau}} + B$$

where A and B are two constants that can be determined using the boundary conditions.

Indeed, for $t = 0$, $v_B = (4V_{CC}/3) = A + B$.

When $t \rightarrow \infty$, $v_B \rightarrow 0 \Rightarrow B = 0$ and $A = (4V_{CC}/3)$

$$v_B = \frac{4}{3}V_{CC}.e^{\frac{-t}{\tau}}$$

When $t = T_1$, $v_B = (V_{CC}/3)$

$$v_B(T_1) = \frac{4}{3}V_{CC}.e^{\frac{-T_1}{\tau}} = \frac{V_{CC}}{3}$$

$$T_1 = 2.\tau.Ln(2)$$

5.1.2) Calculation of the duration T_2 of the low state

The duration T_2 is the time needed for voltage v_B to go from a negative value equal to $(-2V_{CC}/3)$ to a positive value $(+V_{CC}/3)$ following an exponential law:

$$v_B = A_1.e^{\frac{-t}{\tau}} + B_1$$

where A_1 and B_1 are two constants that can be determined using boundary conditions.

For $t = 0$, $v_B = (-2V_{CC}/3) = A_1 + B_1$

When $t \to \infty$, $v_B \to V_{CC} \Rightarrow B_1 = V_{CC}$ and $A_1 = (-5V_{CC}/3)$

$$v_B = V_{CC}(1 - \frac{5}{3}.e^{\frac{-t}{\tau}})$$

When $t = T_2$, $v_B = (V_{CC}/3)$,

$$v_B(T_2) = V_{CC}(1 - \frac{5}{3}.e^{\frac{-T_2}{\tau}}) = \frac{V_{CC}}{3}$$

$$T_2 = \tau.Ln(\frac{5}{2})$$

Finally:

$$T = T_1 + T_2 = \tau.Ln(10)$$

5.2) Expression and value of the cyclic ratio

$$r = \frac{T_1}{T} = \frac{2\tau.Ln(2)}{\tau.Ln(10)} = \frac{2Ln(2)}{Ln(10)} = 0.6$$

5.3) Nature of the output signal

The output signal is a rectangular signal ($T_1 \neq T_2$) since $r \neq 0.5$.

SOLUTION TO EXERCISE 12

1) Truth table of a NOR gate with short-circuited inputs.

Input		Output
1	2	B
0	0	1
1	1	0

It is characteristic of a non-inverting gate.

2) Stage-by-stage operation of the circuit assuming $v_A = 0$ V and evolution of voltages at points A, B and D.

2.1) Stage-by-stage operation

2.1.1) Stage 1

$$v_A = 0 \Rightarrow v_B = V_{CC}$$

$$v_D = 0,$$

Under these conditions, the circuit schematically shown in Figure E12.2 is obtained.

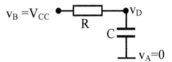

Figure E12.2.

Voltage v_D increases with the charge of capacitor C (charge constant $\tau = RC$)

$$v_D = V_{CC}(1 - e^{\frac{-t}{RC}})$$

As soon as v_D reaches the value $(V_{CC}/3)$ and slightly exceeds it, there is a switching of v_B and v_A.

2.1.2) Stage 2

$v_D = (1/3)V_{CC} + V_{CC} = (4/3)V_{CC}$; $v_B = 0$ and $v_A = V_{CC}$; under these conditions, the circuit shown in Figure E12.3 is obtained.

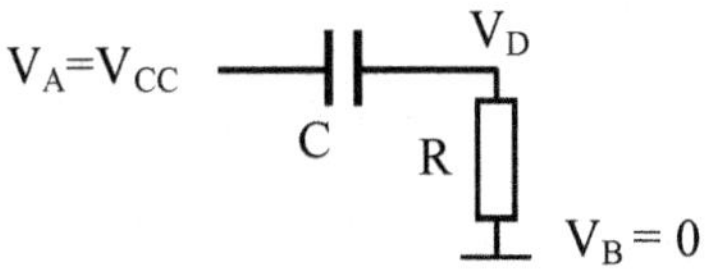

Figure E12.3.

The capacitor tends to bring down to zero the current across it. Consequently, voltage v_D decreases from $(4V_{CC}/3)$.

When v_D reaches $(V_{CC}/3)$ and is slightly below this value, there is a switching of v_B and v_A.

2.1.3) Stage 3

$$v_D = (1/3)V_{CC} - V_{CC} = (-2V_{CC}/3);$$

$$v_B = V_{CC} \text{ and } v_A = 0.$$

Under these conditions, the circuit shown in Figure E12.4 is obtained.

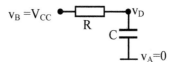

Figure E12.4.

This time voltage v_D increases from $(-2V_{CC}/3)$ seeking to reach V_{CC}, but as soon as it reaches $(V_{CC}/3)$ and slightly exceeds it, there is a switching of v_B and v_A.

2.1.4) Stage 4

$$v_B = 0$$

and $v_A = V_{CC}$, $v_D = (1/3)V_{CC} + V_{CC} = (4V_{CC}/3)$.

This leads back to stage 2. There is therefore an indefinite passage from stage 2 to stage 3 and vice versa.

2.2) Evolution of voltages at points A, B and D (see Figure E12.5)

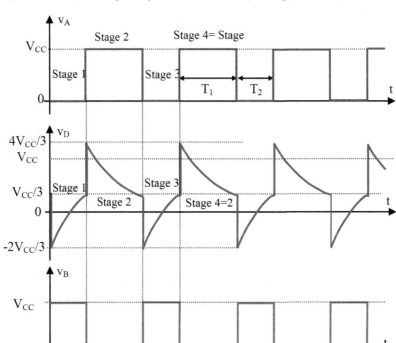

Figure E12.5. *Evolution of voltages* v_A, v_B *and* v_D

3.1) Function of the circuit

Function of the circuit: astable multivibrator

3.2) Calculation of the duration of the high state T_1, the low state T_2 and the oscillation frequency

3.2.1) Duration of the high state T_1

The output is taken at the level of v_A. T_1 is the time needed by voltage v_D to go from the value $(4V_{CC}/3)$ to the value $(V_{CC}/3)$. It should be recalled that voltage v_D evolves exponentially according to the following law:

$$v_D = A_1 e^{\frac{-t}{RC}} + B_1$$

Boundary conditions lead to:

For t = 0, $v_D = (4/3)V_{CC} = A_1 + B_1$ and for $t \to \infty$, $v_D \to 0$: $B_1 = 0$ and $A_1 = (4V_{CC}/3)$

Hence: $v_D = \dfrac{4}{3}V_{CC}e^{\frac{-t}{RC}}$

At t = T_1, $v_D = (1/3)V_{CC}$: $v_D(T_1) = \dfrac{4}{3}V_{CC}e^{\frac{-T1}{RC}} = \dfrac{1}{3}V_{CC}$

$T_1 = 2RC.Ln(2)$.

Numerical application: R = 10 kΩ and C = 1 µF $\Rightarrow T_1$ = 13.86 ms

3.2.2) Duration of the low state T_2

T_2 is the time needed by voltage v_D to go from a value of $(-2V_{CC}/3)$ to $(V_{CC}/3)$.

$v_D = A_2 e^{\frac{-t}{RC}} + B_2$;

When t = 0, $v_D = (-2/3)V_{CC} = A+B$ and for $t \to \infty$, $v_D \to V_{CC}$:

$B_1 = V_{CC}$ and $A_1 = (-5V_{CC}/3)$. $v_D = V_{CC}(1 - \dfrac{5}{3}e^{\frac{-t}{RC}})$

A t = T_2, $v_D = (1/3)V_{CC}$; $v_D(T_2) = V_{CC}(1 - \dfrac{5}{3}e^{\frac{-T_2}{RC}}) = \dfrac{1}{3}V_{CC}$

$T_2 = RCLn(\dfrac{5}{2})$

Numerical application: R = 10 kΩ and C = 1 µF

T_1 = 13.86 ms, T_2 = 9.16 ms; T = $T_1 + T_2$ = RC.Ln10 = 23 ms

3.2.3) Oscillation frequency: F = 1/T = 43.5 Hz.

References

[CAS 66] CASSIGNOL E.J., *Théorie et pratique des circuits à semiconducteurs : électronique non-linéaire*, Dunod, 1966.

[DED 13] DE DIEULEVEULT F., *Electronique radiofréquence : composants pour télécoms, amplificateurs, oscillateurs, PLL, filtres : théorie et simulation : cours et exercices corrigés*, Ellipses, 2013.

[DZI 96] DZIADOWIEK A., LESCURE M., *Fonctions à amplificateurs opérationnels*, Eyrolles, 1996.

[FOU 10] FOURNIOLS J.-Y., ESCRIBA C., *Systèmes électroniques analogiques : amplification, filtrage et optronique*, Presses universitaires du Mirail, 2010.

[GER 05] GERVAIS T., *Electronique, cours et exercices*, Vuibert, 2005.

[HAR 96] HARAOUBIA B., *Electronique générale*, Office des Publications Universitaires (OPU), 1996.

[HER 13] HERMINIO M.G., "On modified Wien-bridge oscillator and a stable oscillator", *Analog Integrated Circuits and Signal Processing*, vol. 75, no. 1, pp. 179–194, April 2013.

[MAL 80] MALVINO A.P., *Principes d'électronique*, McGraw-Hill, 1980.

[MAN 95] MANCINI R., LIES J., Current Feedback Amplifier Theory and Applications, Application note no. 9420, Harris Semiconductor, 1995.

[MIL 79] MILLMAN J., *Microelectronics: Digital and Analog Circuits and Systems*, International Student Edition, McGraw-Hill, 1979.

[MIL 88] MILSANT F., *Cours d'électronique : composants électroniques*, Eyrolles, 1988.

[MOU 70] MOUNIC M., *Semiconducteurs – transistors*, Part 2, Foucher, 1970.

[SEN 16] SENANI R, BHASKAR D.R., SINGH V.K. et al., *Sinusoidal Oscillators and Waveform Generators Using Modern Electronic Circuit Building*, Springer, 2016.

[TIN 11] TINGUY P., Etude et développement d'un oscillateur à quartz integré, PhD thesis, University of Franche-Comté, 2011.

[TRU 09] TRUDGE G., Single transistor crystal oscillator circuits, Application note, RAKON, 2009.

[VAL 94] VALKOV S., *Electronique analogique : cours avec problèmes résolus*, Casteilla, 1994.

Index

Printed in the United States
By Bookmasters